Planting Trees in the Developing World

Steven R. Brechin

Planting Trees in the Developing World

A Sociology of International Organizations

THE JOHNS HOPKINS UNIVERSITY PRESS
Baltimore and London

Printed in the United States of America on acid-free paper

06 05 04 03 02 01 00 99 98 97 5 4 3 2 1

The Johns Hopkins University Press
2715 North Charles Street
Baltimore, Maryland 21218-4319
The Johns Hopkins Press Ltd., London

Published in cooperation with the Center for American Places,
Harrisonburg, Virginia

Library of Congress Cataloging-in-Publication Data will be found
at the end of this book.
A catalog record for this book is available from the British Library.

ISBN 0-8018-5439-3

To Nancy, Maddy, and Archie,
and to all my wonderful students
at Princeton

Contents

Preface

One area that gives grounds for hope, but also requires scrutiny, is the enthusiasm in the 1980s for "social forestry."

JACK WESTOBY, *Introduction to World Forestry.*

Organizations act, but what determines how and when they will act?

JAMES D. THOMPSON, *Organizations in Action.*

If the use of personal interviews, gossip channels, working papers and participation opens the way for errors, it remains, however, the only way in which this type of sociological research can be carried on.

PHILLIP SELZNICK, *TVA and the Grass Roots: A Study of Politics and Organizations.*

With this book, I hope to diminish the sense of mysteriousness that seems to obscure our understanding of the work and behavior of three development organizations: the World Bank, the Food and Agriculture Organization of the United Nations (FAO), and CARE USA. Time and again the international community has turned to these organizations to improve the lives of poor people in poor nations by helping those people to, among other things, better manage their natural resources. Yet, we really know very little about how the three organizations work, nor do we understand why they behave as they do.

The questions I attempt to answer are straightforward. First, why did these three organizations, which happened to have adopted the goal of community forestry almost simultaneously, then respond so differently in striving to achieve that goal? Second, why did their level of success vary so much? As we shall see, each organization has particular strengths and weaknesses, shaped by specific internal characteristics and norms, by its external relationships generally and with other organizations particularly, and by the nature of the community forestry task itself. My basic argument is as follows. Community forestry, as a development assistance innovation, emerged from society's larger political-social-cultural environment, which sociologists like to call the "institutional environment." A number of development organizations, including the World Bank, FAO Forestry Department, and CARE USA, responded and adopted community forestry as a new model of development assistance. The success of their responses, however, varied in part because of the relationship of the organizations' structures with each of their particular technologies, a relationship called "contingency theory" by organizational theorists. In addition, the particular characteristics of their respective core technologies seem to play a critical role in determining these differences. The success of the three organizations was finally shaped by their existence in a more institutional or a more technical environment, or, if you will, by the nature of the external constraints they have each faced, given the community forestry task.

I attempt to answer the book's two central questions—of differing responses and degrees of success—by examining each organization's respective work in community and rural development forestry from the mid-1970s to the early 1990s. The book ends by encouraging movement toward a sociology of international organizations, which I believe might help better prepare us for our future challenges concerning the global biophysical environment and human welfare. In spite of Rwanda and Bosnia, or perhaps in part in response to them, international organizations will likely become increasingly important and more powerful institutions in the decades ahead.

My curiosity about international organizations and their work in development assistance grew quite innocently. While doing fieldwork during the 1980s on reforestation efforts in Niger and Haiti, I was struck by the considerable differences in the activities and effectiveness among the agencies that were sponsoring social forestry projects in those countries. Although my research agenda at the time kept me on the trail of other issues, the question of organizations and their role in generating successful people-related forestry projects eventually piqued my curiosity.

While in Niger, I learned that the United States Agency for International Development (USAID) had been attempting to promote forestry projects over a number of years without much success. The World Bank, in the late 1970s, had established a large fuelwood plantation project that development specialists in Niger considered wholly inappropriate and unsustain able. I also overheard other development specialists discuss the possibility of FAO's establishing a forestry project soon in central Niger, while others replied that they had been hearing that rumor for years. Although the FAO project did emerge several years later, it was the private voluntary organization, CARE USA, and its community forestry work in Niger's Bouza region, that had everyone's attention at the time. Among CARE USA's portfolio was the Majjia Valley Windbreak Project, which has since become a rural development legend in the Sahel. Why did CARE appear to succeed and the others fail in one way or another?

A few years later, in Haiti, a country known for, among other things, its development debacles, USAID was doing much better than it had done in its earlier efforts in Niger. Its "Agroforestry Outreach Project" was reversing Haiti's development tradition. All in all, USAID was administering the most successful rural development effort that had ever taken place in Haiti. A key feature of the project seemed to be USAID's use of several nongovernmental organizations (NGOs) as partners. USAID maintained overall responsibility for the project, but left the day-to-day responsibilities to the NGOs, which appeared better adapted to working directly with local community members on specific forestry-related activities. This organizational arrangement seemed to be an essential component in explaining the project's impressive accomplishments.

Therefore, I decided to do a comparative study of three development organizations that were doing the same kind of work—in this case, community forestry. This kind of approach allows us to consciously highlight the differences among the organizations. A better understanding of what each organization's strengths and weaknesses are, and how the organizations are shaped, may help explain differences in performance in the field. And while I hold out little hope that large, entrenched development organizations will "change their stripes" and transform themselves because of an academic critique, it is quite possible that the kind of information gathered by a comparative study could be used to help them, and the rest of us interested in natural resources management and development, to think about how to improve performance in the field.

I chose the World Bank, FAO, and CARE because they have been quite disparate in organizational structures and philosophies, yet have operated in the common arena of community and rural development forestry. The World Bank is a large multilateral financial and economic development agency. It helps governments invest in development at the national level largely through the financing of specific projects. FAO, which provides professional expertise to member governments on matters of agriculture, for estry, and fisheries, is one of several specialized agencies of the United Nations (UN). CARE USA is a nonprofit, private voluntary organization (PVO) that works more or less directly with people and communities at the local level on development issues and emergency relief.

Since the World Bank, FAO, and CARE have been among the largest and most active aid agencies of their kind in the world, the questions asked in this book will be of practical interest to those involved in forestry as an agent of rural community development, because the answers throw some light on three major "players" on the scene. From my own perspective as a natural resource sociologist interested in organizational behavior, the importance of the answers takes on a more theoretical bent. It happens that the organizational dimensions of development assistance and international natural resource management—and indeed the nature and workings of international organizations in general—have been largely overlooked by scholars. This, to me, is quite a surprise, given the importance these organizations have taken on since the end of World War II.

So the goal in this book is to explore sociologically these general issues by comparing the community and rural development forestry programs of World Bank, FAO, and CARE. In doing so, I draw theoretical inspiration from a number of fields of inquiry, including development studies, natural resource sociology, political science, and organizational sociology. I see the book as a modest effort to start filling a rather large gap among several distinct literatures. In so doing, I have utilized what organizational theorist Richard Scott (1992) calls the "combining the perspectives" approach to the study of organizations (although perhaps a bit more broadly than he intended) (95). And as I suggested above, I hope this book will also be of interest to practitioners. By better understanding the workings and nature of the World Bank, FAO, and CARE, people in the field will be better able to understand how the community and rural development forestry interventions of the three organizations will affect (or fail to affect) what is happening on the ground.

The Term *Environment*

I must say a word or two about language before we go much further. Organizational sociologists like to talk about *environment* too, but this usage is much different from how environmentalists and most others think about the term today. Environmentalists and like-minded people, of course, are referring to nature, natural resources, or ecological systems. Organizational environments, on the other hand, refer to the external forces—relationships with other organizations, government policies, cultural norms and expectations, and a host of other factors—that influence the behavior and structure of a particular organization or groups of organizations. To distinguish between the two meanings of *environment,* I will use *biophysical environment* when I am talking about the state of our world's ecological conditions.

A Note on Methods

Before I acknowledge the great number of people who have assisted me with this project, I should say a bit about the methods employed. The words of Selznick, which were quoted in the front of this preface, ring true here. This book is based on qualitative research, using primary and secondary sources such as published literature, organizational documents, and other publications, complemented with site visits and a large number of personal interviews.

Many of the concepts I discuss are far from straightforward and clearcut. Rather, they are messy and often elusive, such as the constructs of core technologies and performance. Still, I strongly believe that to avoid the concepts because they are complicated or awkward would only cause us to avoid topics of great importance. I am also amazed at how little information exists on the organizations themselves. With the important exception of the World Bank, very few accounts, good or otherwise, exist on CARE USA and especially on FAO Forestry. For these organizations, I had to depend more on internal documents and interviews than I would have liked.

Research for this book began in 1987 and ended in 1995. The information of specific forestry projects was derived from internal data dating back to the first forestry program that could be identified for each of the three organizations. In the case of the World Bank, this was the 1960s; for CARE, the mid-1970s. For FAO Forestry, however, the large volume of forestry project data, combined with poor records before 1980, forced me to limit my data collection to the period 1980–90. Forestry project data were extracted from organizational documents. For the World Bank, I used a combination of its annual reports and an internal forestry-related document, the "Review of the World Bank Financed Forestry Activity," produced by the World Bank's forestry advisor's office. I started the collection process before the World Bank's Operations Evaluations Department (OED) made its own review (World Bank 1991b), which provided a list of projects. Careful observers will note discrepancies between my list and the OED's. Although differences or minor omissions will have little, if any, effect on my analysis, it may be of interest to note that World Bank staff were not in the least surprised to learn of those differences. The staff went on to indicate that they had no particular faith that their list was any more accurate than mine, given the poor documentation the World Bank maintains on its former projects. Forestry project data for FAO came from the "Forestry Department Catalogue" (1980–90), which is published once or twice a year by FAO's Operations Service. CARE USA's forestry project census data were compiled from project budget summary forms, called APD-2s, for the fiscal years 1976–92. Additional data were obtained from various project documents in the CARE office in New York (it moved to Atlanta, Georgia, in 1994). A simple coding scheme was devised for placing specific projects into forestry project categories, traditional forestry projects, community and rural development forestry projects, and a residual category of "mixed forestry projects," which frequently captured larger projects that contained elements of both traditional and rural development. I have provided a more detailed description of the coding process and other methodological considerations, along with the actual project data (Brechin 1994).

I made a number of trips to Rome, to Washington, D.C., and to New York City. In all, over seventy individuals were interviewed in person or by letter, e-mail, or phone; I questioned many key informants more than once. In an effort to improve validity, findings I considered crucial to my arguments were corroborated by at least three sources.

Acknowledgments

A number of individuals and organizations have been instrumental in making this book a reality and need to be acknowledged. Staff at the three organizations were extremely important in my efforts. Especially so were John-Michael Kramer, Nina Bhatt, Margaret Ford (archivist), Tim Aston, Frank Brechin, Peter Hazlewood, Tom Painter, and Marshall Burke, all of CARE USA. At FAO Forestry, many of the staff were extremely helpful. A special note of thanks goes to Marilyn Hoskins for hosting my visits to Rome and for arranging many of my meetings. At the World Bank, William Beattie, John Spears, Chip Rowe, and William Magrath were invaluable for their insights and for providing contacts. Richard Pardo was extremely helpful in a number of capacities.

Funding for this book came from several sources. The University of Michigan and its Department of Sociology, Rackham School of Graduate Studies, and School of Natural Resources and the Environment helped to move this project forward during its earliest phases. More recently, generous support was provided at Princeton University from the following: the Center for Energy and Environmental Studies, the Tuck Fund from the Dean of the Faculty's Office, and the Center of International Studies and its Pzifer Fund. Most of the later funds went to pulling together and analyzing the enormous amount of forestry development data collected for this project. Brad Diamond and, especially, Terence Kelly provided invaluable technical assistance in helping to organize the large amount of project data that laid the foundation for the discussion found here. Donna DeFrancisco graciously provided some typing assistance for a number of the tables and figures. Casey Crawford helped pull together the bibliography.

Many individuals provided intellectual support for this project. They included Gayl Ness and Pat West at the University of Michigan and Gene Burns, Paul Clements, Marc Levy, Michael Ross, and, particularly, Frank Dobbin of Princeton University. Viviana Zelizer, Marvin Bressler, Robert Socolow, Hal Feiveson, and John Waterbury, also at Princeton, gave encouragement that pushed this project forward in a number of ways. A very special thanks goes to my friend and colleague David Harmon (of the George Wright Society) for improving this manuscript markedly with his wit, insights, and red pen. I owe much to George F. Thompson, president of the Center for American Places and my editor for the Johns Hopkins University Press, and Carol Mishler, associate editor at the Center, for their enthusiasm and support for the project, as well as for helping to shape an incoherent manuscript into, I hope, a useful book. I must acknowledge, too, the useful comments from the reviewers for the Johns Hopkins University Press. Of course, all inaccuracies, follies, and other imperfections are solely my own. Finally, Nancy Cantor and our children, Maddy and Archie, as well as the rest of our families, require special awards for all their love and support and for the considerable patience that they displayed much too often amidst the stresses and strains of my researching and writing this book.

Acronyms

CARE	Cooperative for American Relief Everywhere (a PVO)
CIDA	Canadian International Development Agency
COFO	Committee on Forestry (FAO Forestry's representative to the FAO Council)
CSD	United Nations Commission on Sustainable Development
DANIDA	Danish International Development Agency
ECE	United Nations Economic Commission for Europe
EMENA	The region of Europe, Middle East, and North Africa
FAO	The Food and Agriculture Organization of the United Nations (a technical agency)
FLCD	Forestry for Local Community Development (an FAO forestry program)
FTPP	Forests, Trees, and People Programme (an FAO Forestry program)
IBRD	International Bank for Reconstruction and Development (the World Bank)
IDA	International Development Association (Soft Loan Affiliate of the World Bank)
IGOs	International Governmental Organizations
IMF	International Monetary Fund
INGOs	International Nongovernmental Organizations
IPCC	Intergovernmental Panel of Climate Change
IUCN	The World Conservation Union (formally the International Union for Conservation of Nature and Natural Resources)
LAC	The region of Latin America and the Caribbean
NGOs	Nongovernmental Organizations
PCVs	Peace Corps Volunteers
PICOP	Paper Industries Company of the Philippines
PVOs	Private Voluntary Organizations
SIDA	Swedish International Development Authority
TFAP	Tropical Forestry Action Programme (an interorganizational effort)
UNCED	United Nations Conference on Environment and Development
UNDP	United Nations Development Programme
UNESCO	United Nations Educational, Scientific, and Cultural Organization
UNHCR	United Nations High Commissioner for Refugees
USAID	United States Agency for International Development
WCARRD	FAO's World Conference on Agrarian Reform and Rural Development
WHO	World Health Organization
WRI	World Resources Institute (an environmental NGO)

Planting Trees in the Developing World

Introduction

Organizations are among the most powerful structures humanity has yet created in an effort to serve its ever-expanding needs. They can also be among the most deceitful. Almost always established with good intentions, organizations are blessed with skills to acquire, arrange, and transform scarce resources into products and services that are thought useful to the larger society. Certainly international organizations have been viewed that way. With the exception of the World Bank, which has lately received much negative attention for its activities, people assume that an organization does what it was created to do, can perform a variety of task well, or can do so without undue costs. Some people may wonder how an organization can mobilize its resources only to accomplish little or nothing—or, worse yet, the opposite of what was intended. Most people may shrug off these episodes as simply bungling or as failures arising from an overly difficult task. Few people give much thought to the possibility that an organization could have been inappropriately designed to carry out certain tasks, that it could have fallen under the sway of powerful interest groups, or that it could purposefully deflect the demands of its creators to pursue its own bureaucratic self-interest.

The point of this book is not to dwell on the sinister possibilities of organizations but to encourage all of us to think more critically about organizations and to ensure that they work for our collective well-being—or at least to ensure that we have realistic expectations of them. It is also the purpose of this book to introduce some of the literature and perspectives of organizational sociology to students and practitioners who have interests in environment and development.

One key point, however, is to explore the possible strengths and weaknesses that each organization possesses. In this volume, I attempt to identify the strengths and weaknesses of three organizations—the World Bank, FAO, and CARE USA—as each organization has worked on rural development forestry from the mid-1970s to the early 1990s. My analysis suggests that each organization harbors a unique set of strengths and limitations when it comes to this specific forestry task. Such a finding may be helpful when fostering success in this and similar types of development work. Near the end of this book, I make a few policy suggestions that might lead to greater organizational performance generally and that might improve the delivery of "bottom-up" rural development efforts more specifically.

At a more theoretical level, I urge the development of a sociology of international organizations. Like Easterbrook (1995), I believe that the mounting threats to the global environment and to human welfare will cause us to turn increasingly to international organizations for the management of our welfare. To date, little sociological attention has been focused on these types of organizations, how they work, and the forces that shape them.

To begin pushing this agenda forward, I combine a number of organizational theories, including institutional, contingency, resource dependency, and core technologies. In analyzing the three organizations, one can find support for each of these perspectives, but some theories carry greater powers of explanation than others. For example, characteristics of the technical core seem to carry more weight for discussions of the World Bank, resource dependency for FAO, and contingency theory for CARE USA, whereas institutional theory applies to all of them.[1] Still, each of these notions, taken alone, comes up short. The findings support not one perspective but several simultaneously. Only by merging perspectives does one get a more complete theoretical "fix" on these organizations, as was suggested by Thompson (1967) three decades ago and is emphasized by Scott (1992) today.

The pivotal merger of this study is institutional theory with contingency theory. Like Gupta, Dirsmith, and Fogarty (1994), I have combined the insights of these two important theories to understand the behavior and performance of organizations.[2] I draw upon institutional theories, uniting those of sociology (DiMaggio and Powell 1991; Strang and Meyer 1993) with those of political science (Hall 1993) to understand more fully the diffusion and adoption of community forestry practices among and by the three organizations, representing a shift in "policy paradigms" of economic development models. To better grasp the difference in their actual performances, however, we needed to bring in contingency theory.

When I began this study, I expected to find that the three organizations began their programs in rural development forestry at markedly different dates. As we shall see, the time differences were trivial. Clearly, isomorphic processes of new institutionalism, or the adoption of identical structures among different organizations (DiMaggio and Powell 1991), were at work in creating three similar rural development programs in three very different organizations at more or less the same time. Forces within the macrolevel, or overarching, environment of these organizations were identified fairly quickly, which explained the diffusion of this new kind of forestry. However, identifying the forces of institutionalization did little to explain the striking differences among the organizations' programs. For such an explanation, I turned to contingency theory and those perspectives more allied with the "old" institutionalism of the natural system model, or the notion that organizations are composed of competing interests and goals.

Drawing upon arguments from contingency theory, one of the book's central premises is that if an organization is to be successful at a given task, there needs to be a good "match" among its internal elements, the external forces affecting its output, and the task to be achieved. Chronic poor performance on a given task suggests a mismatch of some kind. In some ways the creation of detached structures, or the establishment by executives of a unit not really vital to the organization's mission, as suggested by Meyer and Scott (1983) and others, can be seen as nothing more then a type of politically motivated coping strategy for overcoming poor performance that results from a forced mismatch through the manipulation of influential institutional symbols (see Brechin forthcoming). Defining performance, however, which is central to contingency theory, turns out to be more difficult than originally thought. The issue is vastly complicated by the many social constructions created by various parties who have their own notions and self-interests.

Overview

In the rest of this chapter, I summarize the book's findings and arguments and conclude by discussing the theoretical foundations for this work as well as defining key concepts. In Chapters 1, 2, and 3, I review the organizational elements of the World Bank, FAO, and CARE, respectively. In particular, I trace the historical development of these organizations and the creation of their respective organizational characters and present the features of their core technologies, task environment, and staffing and structures. These analyses lay the foundation for our study of the work of these organizations in community and rural development forestry.

In Chapter 4, I answer a number of questions about the forestry programs of the Bank, FAO, and CARE: namely, *when, how, how much, where, how well,* and, *why. When* and *how* the organizations became involved in rural development forestry are discussed under the heading "Program Entrance," where I review not only the dates when the first rural development forestry projects were established but also the formal structures to administer those projects. In addition, I consider how each organization came to be involved in community forestry. The question of *how much* work has been done in this area is taken up under "Program Commitment" by looking at the organizations' project expenditures across time. The question of *where* projects have been promoted is summarized under "Program Appropriateness." Appropriateness is assessed by reviewing whether or not projects were established in countries and regions that were recognized as being in great need of rural development forestry interventions. *How well* the organizations' programs worked is discussed under "Program Performance." Performance was originally defined as the ability of the organizations to reach and benefit the rural poor. Although important, this definition perhaps is too narrow in representing positive outcomes. The question of multiple definitions of performance is raised throughout this study.

Chapter 5 summarizes the findings about the organizations, their characteristics, and their work in more conceptual terms. The chapter concludes with several policy-related suggestions. Finally, for interested readers, Chapter 6 continues the more theoretical discussion of the results, drawing upon—and I hope, adding to—ideas and findings from organizational sociology with the aim of moving toward a sociology of international organizations.

A Comparative Analysis of Rural Development Forestry Programs: Summary of Findings

Table I.1 presents the study's overall findings. As noted above, the data are organized by four principal concerns: program entrance, commitment, appropriateness, and performance.

Program Entrance

Two findings emerged. First, the dates when these organizations started their work in community forestry differ only slightly. The organizations established aspects of their programs within three years of each other. Processes of isomorphism (DiMaggio and Powell 1983, 1991), wherein organizations are shaped by broader cultural influences from larger society, were clearly at work here. Second, in the end, learning about *how* these organizations adopted these forestry innovations turns out to be more insightful

Table I.1
Summary of Findings

	Program Entrance	*Level of Commitment*	*Level of Appropriateness*	*Level of Performance*
FAO Forestry	Mixed*	Low to Moderate	Moderate	Variable†
World Bank	Moderate	Low to Moderate	Low	Low†
CARE Int.	Mixed*	High	High	High†

*The evidence conflicted, but overall differences were slight.
†There are important exceptions to these generalties.

than knowing *when.* Drawing upon Strang and Meyer's (1993) theoretical contributions on institutional adoption was particularly helpful. We continue by reviewing when each organization adopted its rural development forestry programs.

At first glance, the World Bank appears to be the first of the three organizations to have sponsored rural development forestry projects, beginning in 1974 with a loan for the Paper Industries Company of the Philippines (PICOP) project. But, as we shall see, the PICOP forestry loan was an outlier. The Bank did not become consciously active in community and rural development forestry until it published its first, very influential forestry policy statement (World Bank 1978).

I had expected to find that CARE led the movement into this new type of forestry, but its first serious projects did not get under way until 1975 (fiscal year 1976), a year after the Bank's PICOP loan. More problematical in establishing a start date was the fact that CARE was the last of the three organizations to create a formal structure to promote its forestry-related operations. Not until 1981, when CARE and USAID signed their first program matching grant for forestry, did CARE create its Renewable Natural Resources position.

Instead of being the laggard—which many critics might have expected, given its reputation as a slow-moving bureaucracy—FAO Forestry was right there with CARE and the Bank in promoting community forestry through its Regular Programme activities carried out with the support of the Swedish International Development Authority (SIDA) in the mid-1970s. And, as the analysis in Chapter 2, shows FAO Forestry was certainly involved with rural development forestry projects in the field by 1980, when our data for this organization begins. With the 1978 World Forestry Congress in Jakarta, FAO Forestry was also instrumental in helping to establish an international consensus on the importance of community forestry. Other structures in FAO Forestry quickly followed, with SIDA formally establishing the Special Programme for Forestry for Local Community Development (FLCD) in 1979. Immediately after the FAO sponsored the World Conference on Agrarian Reform and Rural Development (WCARRD) in 1980, FAO Forestry established its Forestry for Rural Development Technical Programme in the same year.

While investigating when these three organizations became involved, two insights emerged. First, both indicators used for this analysis—the date of the first rural development field project and the date of the establishment of formal organizational structure—conflicted only slightly with one another. Second, both sets of dates shed little light on the events surrounding the organizational movement into this new type of forestry. For example, the Bank's 1974 PICOP loan, its inaugural community forestry project, was essentially a spurious event, as already noted. The PICOP project was originally neither designed nor promoted by the Bank. It was already a successfully established pilot project, created by the paper company when it was referred to the Bank for additional capital expansion by the Development Bank of the Philippines. That is, PICOP fell into the Bank's lap.

On the other hand, CARE and FAO Forestry's start dates in rural development forestry, 1975 and 1976, respectively, look much more substantive. Their first projects became the foundation for rapidly increasing involvement. The Bank's base point for sustained activities did not begin until 1978. Table I.2 contains the program entrance dates for each of these organizations. As the table shows, the outcome depends upon which evidence was presented, but the differences are trivial.

As it turns out, it is more illuminating to look at *how* these organizations became involved in rural development forestry. The adoption of community forestry practices by these organizations, however, is discussed in some detail in Chapter 6.

Program Commitment

Program commitment reflects the amount of resources and activities each organization has devoted to community and rural development forestry. Here we are concerned with the percentages of financial resources and number of projects attached to rural development forestry as compared with (1) the total forestry activities of each organization and (2) the total value of each organization's work in development during the base year of fiscal

Table I.2
Comparison of Program Entrance into Rural Development Forestry

	Date of First Project	*Date of Formal Structure/Policy*	*Name of Structure/Policy*
CARE USA	1975	1981	RNR Coordinator's Position
FAO Forestry	1976*	1978	World Forestry Congress
		1979	FLCD Program Established
		1980	Forestry Rural Development Program
World Bank	1978†	1978	Forestry Sector Policy Paper

*These are Regular Programme activities. The date of first Field Programme rural development forestry project is not known since project data begins in 1980.
†After removing 1974 outlier.

year 1992. Table I.3 (a and b) presents the comparative data. CARE's high commitment to rural development forestry was explained wholly by the fact that it is the only type of forestry it practices.

Although all of CARE's forestry work was allocated to community forestry, FAO Forestry sponsored more projects than CARE (189 to 96) in more countries (71 to 29), far more than the World Bank. FAO Forestry's greater breadth of activities stemmed largely from its legitimacy as the international community's forestry agency, with its sole "business" being forestry. The high project and country numbers were tempered, however, by the fact that only a comparatively small percentage of its forestry programming (25%) found its way to community and rural development concerns. Most of FAO's allocations to forestry activities were geared toward traditional forestry. This turns out to be less the fault of FAO Forestry itself and more the fault of its financial supporters.

The Bank, with its vast financial resources, provided considerably more money (1.4 billion 1992 dollars as loans) for just community and rural development forestry than did the other two organizations combined ($300 million). That amount was about 41 percent of the Bank's total work in forestry. In addition, about 42 percent (or 36 of 85) of the Bank's forestry projects were for community and rural development forestry. Moreover, these projects were located in only about 39 percent of the countries (18 of 46) where the Bank sponsored forestry projects. In particular, India dominated the Bank's work in community and rural development forestry, especially in the 1980s. By 1992 India alone had eight community forestry projects with a combined value of $470 million (nominal), or 41.3 percent of the

Table I.3
Comparison of Program Commitment:
Community and Rural Development Forestry

(a)
Comparison with Total Forestry Activities

	Amount (U.S. millions)				
	Nominal	*Adjusted (1992)*	*% of Amount to Total Forestry*	*No. of Countries/ Total Number (%)*	*No. of Projects/ Total Number (%)*
World Bank	1137*	1466*	41	18/46 (39)	36/85 (42)
FAO Forestry	142†	184†	25	71/112 (64)	189/691 (27)
CARE USA	104‡	123‡	100	29/29 (100)	96/96 (100)

*Amount of Approved Loans 1969–92.
†Amount of Approved Budgets 1980–90.
‡Amount of Approved Budgets 1975–92.

(b)
Comparison with Total Development Activities (1992)

	1992 Amount of Rural Development Forestry (U.S. millions)	*1992 Amount of Total Development Activities (U.S. millions)*	*% Rural Development Forestry of Total Development Activities*
World Bank	331.20	21,705.7	1.52
FAO Forestry	25.02*	2,533.3	.99
CARE USA	18.10	242.9	7.45

Source: Brechin, Forestry Project Data for the World Bank, FAO, and Care International, 1969–1992 (1994, mimeograph); World Bank, *Annual Report* (1992); CARE USA, *Annual Report* (1992); FAO, *Annual Review* (1992).
*1990 total, compared to all FAO field activities.

Bank's total work in that type of forestry. Without India, the Bank's activities in this new forestry would have been considerably less.

Overall, however, forestry remains a relatively small area of work for all three organizations. In 1992, for example, forestry as a sector (i.e., all types of forestry projects) composed only 1.5 percent, 0.99 percent, and 7.45 percent of the development assistance work for the Bank, FAO, and CARE, respectively (see table I.3b).

Table I.4
Comparison of Geographic Distribution:
Community and Rural Development Forestry Projects
(% of Total Financial Value of Rural Development Forestry)

	World Regions					
	Asia	*Africa*	*Lac*	*Emena*	*Global*	*Total*
World Bank*	51.3	18.7	21.3	8.7	—	100.0
FAO Forestry†	19.6	53.9	19.8	4.3	2.4	100.0
CARE USA‡	11.8	61.5	26.3	0.4	—	100.0

*Amount of approved loans, 1969–92 (rural development forestry).
†Amount of approved budgets, 1980–90 (rural development forestry).
‡Amount of approved budgets, 1975–92 (rural development forestry).

Program Appropriateness

Table I.4 shows the distribution by project value of the three organizations' work in community and rural development forestry, 1969–92. Three observations stand out. The first is the uneven distribution of the Bank's work. Fifty-one percent of the value of the Bank's work was in Asia, just less than 19 percent in Africa, and about 30 percent in the other world regions. This is in sharp contrast to the uneven distribution of CARE USA's work: 61.5 percent in Africa, and 26.3 percent in Latin America, mostly Central America. The other regions were more or less ignored. FAO Forestry's work is a bit more evenly distributed than that of the other two organizations, but with an emphasis on Africa over Asia and Latin America. And finally, reflecting its international stature as the world's forestry agency, only FAO Forestry works on global and interregional projects. Although interesting, the actual distribution of work is most meaningful when it can be compared to actual human need.

To more accurately evaluate the appropriateness of each organization's program, I have compared the geographic distribution of project activities with a list of countries that have suffered severely from the consequences of deforestation and land degradation. The points of comparison are from a 1980 study by FAO Forestry on worldwide fuelwood scarcity and deficits and are used here as an indicator of international need in community and rural development forestry. The project activities were compared in those countries that relied on fuelwood for at least two-thirds of their energy and that were already suffering from acute fuelwood scarcity.

Regarding program appropriateness, the Bank fared rather poorly in

Table I.5
Comparison of Program Appropriateness:
Program Activities in Countries
Dependent on Fuelwood for Two-Thirds of Energy Consumption
(Rural Development Forestry to Total Forestry Activities)

	$ Amount (Millions)	*No. of Projects*	*No. Countries*	*% Rural Forestry Program $*	*% of Total Forestry $*
	1992 dollars				
World Bank	369.0	15	8/12	24.8	<10.0
FAO Forestry	86.3	66	10/12	47.0	12.0
CARE USA	75.5	52	11/12	61.5	61.5

most categories, save one: the dollar amount of the financial investment. Table I.5 presents these comparative findings.

Although the Bank made loans of nearly $369 million to the most desperate countries, which was more than double the combined value of expenditures of CARE and FAO, only fifteen projects were located in 8 of the 12 critical countries on that list. Only 25 percent of the Bank's activities in community and rural development forestry were located in those countries, which composed less than 10 percent of the Bank's total forestry activities.

FAO Forestry fared better, with 66 projects in 10 of the 12 countries, representing 47 percent of its community and rural development forestry activities. CARE USA's program was even more appropriate. It sponsored 52 projects in 11 of the 12 countries for $75.5 million, that is, 61.5 percent of its forestry program went to these most desperate countries.

Program Performance

Table I.1 also presents the summary of the findings on how well the three organizations did in directing their projects to the needy rural poor. Generally speaking, the Bank performed poorly, FAO Forestry moderately, and CARE very well. As is noted in each of the case study chapters, however, the concept of performance turned out not to be as straightforward as expected. A summary of the findings is given in Table I.6.

The Bank's work in this new forestry did not fare well, especially in the achievement of the technical objectives for these types of projects. Its projects in Africa, for example, were especially inappropriate and poorly designed. Most project benefits in India as well as Africa accrued largely to

Table I.6
Comparison of Performance

WORLD BANK

- Possessed inappropriate or poor technical project designs.
- Project benefits were frequently not directed to rural poor.
- Recipient governments frequently lacked ability to implement projects.
- Performed well when providing expansion capital to established project.
- Stimulated international interest with policy papers and funding.
- Possessed ability to work on macrolevel policy issues.

FAO FORESTRY

- Had highly variable performance level in Field Programme activities.
- Performed best when donor agency and host government were supportive of FAO Forestry's technical efforts.
- Competence and diplomatic skills of FAO field forester were critical.
- Had high performance in Regular Programme activities.
- Possessed professional ability and legitimacy to work with government agencies on macrolevel policy and institutional issues.

CARE USA

- Had technically sound and appropriately designed, decentralized community projects.
- Had efficient and effective projects.
- Project benefits were frequently directed to rural poor.
- Was a superb program and project innovator.
- Had questionable effectiveness with landless and women.
- Had very limitd impact on macrolevel, institutional setting.

government, urban dwellers, or the rural rich. The rural poor generally received little of the advantages. Many of the Bank's rural development forestry projects became new forms of industrial forestry.

The Bank performed well in other capacities. It was successful in stimulating interest in the concept of community and rural development forestry through the publication of its forestry sector policy paper in 1978. From 1969 to 1992 it moved more than $1.4 billion in the name of this new type of forestry, far more than any of the other organizations. The Bank's more successful projects, such as the PICOP in the Philippines and its community forestry project in Nepal, were the result of the Bank's providing needed capital to further the development of already-established and effective projects.

FAO Forestry's performance was highly variable. Beyond the specter of its involvement in the controversial Tropical Forestry Action Programme (TFAP),[3] the work of its Regular Programme, or its headquarters operations, generally received high marks. Its Field Programme, or overseas technical assistance operations, generated less praise but did perform well in some circumstances. As mentioned in the final chapter, supportive circumstances emerged when the tripartite arrangement of donor agency, recipient government, and FAO Forestry collectively rallied around the technical objectives of the project. The technical competence and diplomatic skills of the FAO professional were often key to project success. Although a few donor governments sometimes placed minor conditions on their funding of community forestry projects, that was something rarely done by the United Nations Development Programme (UNDP), FAO's major funder of projects. Finally, FAO Forestry had the professional capabilities—but most important, the political legitimacy—to work directly with government agencies and officials and on national issues, resulting in a greater influence on the macrolevel conditions of forestry efforts within given countries and the world community.

CARE received generally high marks for its field projects. Its projects tended to be appropriate and well designed, efficiently run, and effective. Although its skills and sensitivities for promoting true bottom-up community involvement were frequently questioned, as was its ability to work successfully with the landless and women, poor rural communities were still largely the major beneficiaries of its projects. This was the result of CARE's ability to work directly with rural communities in a more decentralized fashion, while at the same time controlling the project's resources and guiding its operations.

In spite of CARE's generally good technical achievements, the overall impact of CARE's work was questionable. Of the three organizations, it had the lowest total value of work in rural development forestry from 1974 through 1992 ($123 million in 1992 dollars), and this included considerable support from USAID. Its projects were small compared with those of the other organizations and were often situated in larger institutional and political settings that were counterproductive to the projects' overall missions. Although CARE performed well as a project and technical innovator, it lacked the ability and legitimacy to directly influence the larger political setting where it operated, thereby reducing the effectiveness of its efforts.

These findings raise a number of questions beyond those I have already posed: Why was CARE's commitment to rural development forestry, regarding percentage of resources allocated, relatively high, while FAO Forestry's was low? Why did the Bank focus its work mostly in South Asia, and

CARE in Africa? How can we explain CARE's directing most of its efforts to the world's neediest countries while the Bank did very little for those countries? Why did FAO Forestry's Regular Programme, overall, fare better than its Field Programme? Why, even though CARE tended to achieve the technical objectives of its projects, were its effects on the macrolevel settings essentially nil, unlike those of the Bank and FAO? Although committed to the goals of rural development forestry, why did the Bank's projects tend to become new forms of industrial forestry, benefiting everyone but the rural poor?

As will become clearer, each of the organizations used very different strategies, or "policy instruments," in promoting the rural development goal. And the instruments were determined by mixing the internal and external characteristics and forces of each organization. The particulars of these unique strategies created the difference in outcomes.

Theoretical Arguments and Foundation

These findings and the questions they spawn can be explained, I believe, by the following theoretical argument. Community forestry, as a new innovation in rural development, emerged from the larger institutional environment of the development assistance community. The need for this innovation grew from long-term empirical observations of continuing problems with incremental change within the urban-based industrial development approach. In particular, the greater impoverishment and isolation of essentially agricultural regions called for more large-scale shifts in development programming to revitalize them, very much like the process outlined by Peter Hall's (1993) discussion of shifts in policy paradigms.

As this new development need became accepted and established, development agencies attempted to respond to the needs articulated by the institutional environment of which they were a part by adopting rural development programming, including projects from the forestry sector. Because this innovation was identified and sanctioned by forces within the development community's collective culture as a model of progressive policy, such organizational players as the three organizations and a host of others, like bilateral development aid agencies, adopted this new idea at about the same time, collectively helping to legitimize and institutionalize the idea.

Although this isomorphic process of institutional theory is helpful in explaining why these organizations began their programs at the same time, it does not provide much insight into why the organizations responded so differently in their efforts and level of performance, given the same organizational task. I argue that the nature as well as the success of their responses varied in part because of "contingency theory," which refers to the type of relationship the organization's structure has with the characteristics of its organizational technology and with its environmental setting. As part of contingency theory, I also pay close attention to the nature of each organization's core technology, or its ability to transform inputs into outputs. This ability, I believe, has an important defining effect on organizational behavior. The behavior and outcomes of the three organizations were also dependent on whether or not they were embedded in more technical or institutional environments. These two different types of environment reflect to a greater or lesser extent market and nonmarketlike conditions. Organizations in more technical environments worry more about competition and the efficient production of their outputs in order to survive. For those organizations in more institutional environments, the worry is essentially about form over substance. In sum, the behavior of these organizations is determined by a unique and complex set of interactions between each organization's internal characteristics and external relationships. To more fully understand these organizations and their behavior requires the mixing of theoretical arguments and perspectives from within organizational sociology and elsewhere.

The theoretical foundation for this work is constructed from a number of disciplinary-based fields, including organizational sociology, development studies, natural resource sociology, and political science. Some effort is directed at building upon several notions of institutional theory. This is discussed more in Chapter 6.

Organizational Sociology

Much of the earlier sociological work on organizations was directed at finding universals among those organizations. James D. Thompson, in *Organizations in Action* (1967), struck out in a different direction by attempting to explain the differences between organizations. He suggested that variations in organizational behavior and performance were determined by the numerous, uncertain, and complex interactions among the internal and external elements of organizations: technology, structure, and environment. The interactions created unique sets of strengths and constraints for each organization, making each distinctive.

In a comparative study such as mine, one ought not be satisfied with simply reciting development failures—which is easy enough to do—but rather should explain *why* things did not work. Moreover, one ought to try to explain the successes, too, as well as the vast majority of efforts that fall somewhere in between. One way to do this is to think of an organization as a tool, a device specifically designed to carry out a particular task. Obviously, tools are tailored for certain jobs but not for others. In a pinch one could, for example, drive a nail with a pipe wrench and perhaps do it as well as if one had used a hammer. But one would not want to build a whole house that way. Similarly, an organization could conceivably carry out a task that is outside its "natural" strengths, given its current configuration, but not as efficiently or as repeatedly as an organization that is better tailored to that task. I argue strongly that success depends in large part upon whether or not there exists a match between the particular organization and the particular task.

Of course, we should not let the tool analogy carry us too far from the realization that organizations are complex human institutions, not mere mechanisms. As Phil Selznick has noted, organizations are tools that become infused with value. They take on lives of their own. They emerge and develop, honing their inherent strengths, while usually exacerbating or attempting to cover up their weaknesses. These organizational features are defined by complex and continual interactions among an organization's internal characteristics and external (or environmental) influences, at both micro-, or local, levels (i.e., around the precise task to be performed) and macro-, or larger, societal levels. Sociologists Paul DiMaggio and Walter Powell (1991) call these two types of organizational influences "old" and "new" institutionalism, respectively.[4] I try to integrate the old and new institutionalism of sociology (along with the institutionalism of political science) with more technical perspectives to better explain organizational behavior and performance. In particular, to achieve a new policy goal (Hall 1993) of community forestry, new institutionalism rightly focuses our attention on the creation of isomorphic organizational structures already found in society, but the similarities may be only skin-deep, as I found here. And understanding the differences between the outcomes of these similar-looking structures requires that we return to the organizational characteristics that constitute old institutionalism. To fully assess organizational behavior, one must take into account as well (though often to a lesser but still important degree) the distinctive internal elements of each organization, including the characteristics of its structure and core technology, known as contingency theory.

I assert later that performance can best be enhanced, although never guaranteed, when organizations with different strengths work in concert to complement one another when undertaking particular complex tasks, that is, when each is limited only to its particular strengths for the task at hand. Here it becomes necessary to talk about working out specific arrangements between differing organizations and organizational fields. These arrangements too, however, are likely difficult to negotiate and maintain.

Development Studies

Little work on the organizational dimensions of development agencies themselves has been completed since Judith Tendler's classic 1975 study entitled *Inside Foreign Aid.*[5] Drawing upon organizational sociology, Tendler showed how the behavior of development organizations was affected by their structures and environments. In a finding that was surprising for its time, Tendler concluded that there were similar behaviors among bilateral and multilateral agencies. This conclusion was striking since multilateral agencies were free of the kind of political constraints thought to have been hindering the effectiveness of bilateral development programs. Of course, there is more to the study of development organizations than the work of Tendler. Still, most of the studies of organizations have lacked both sociological imagination and focus on organizational sociology, particularly when dealing with the large development organizations.

I have adopted a number of Tendler's perspectives in exploring the activities and behavior of the World Bank, FAO, and CARE. Some of the terms and ideas will be similar, such as the "fit" between the task to be performed and the organization, as well as the role of "task environment" in shaping the behavior of the organization. These similarities are intriguing, I think, for a couple of reasons. First, a number of Tendler's insights continue today as persistent qualities of development organizations. Second, based on these qualities, there may be a need to more adequately institutionalize our knowledge about the nature of development organizations themselves if we are to promote better development.

Still, this effort differs from Tendler's in several basic ways. It does not compare and contrast bilateral with multilateral agencies in an effort to determine the relative advantages of one over the other. Instead, it compares a multilateral assistance agency (the Bank), a hybridized consultative and assistance agency (FAO), and a large private voluntary organization (CARE) in an effort to find relative strengths and weaknesses within each of these organizational types.

Whereas Tendler looked at development assistance generally, this book focuses on forestry-related rural development projects particularly. Both books draw upon organizational sociology but have somewhat different theoretical perspectives. Instead of focusing simply on task environment, I pay some attention as well to the interaction between the more internal technical processes of the organizations and another concept of organizational environments—new institutionalism, as I have already noted—in the hope of better enhancing our understanding of the development process, of development organizations per se, and of organizations more generally.

Unlike Tendler, I examine these three organizations as an outsider, having never been employed by any of them, which has its disadvantages in organizational analysis. This undoubtedly has hindered me from obtaining a more intimate understanding of the organizations. But at the same time my lack of intimacy has allowed me to examine these organizations from what I hope is a more detached and objective vantage point, a position that has allowed me to ask the seemingly "silly" and "naive" questions that may in fact have produced illuminating answers.

Although my effort focuses on the rural development forestry activities of only three organizations, the general themes and lessons here can be applied to a wider range of development organizations and activities, as well as to organizations in general. The more specific findings, however, are likely to be limited to the three organizations and their work in "bottom-up" rural development. Still, it is hoped that this book will stimulate greater thought and research on how development agencies themselves affect the processes of economic development in our world's poorer regions, as well as (more generally) on why organizations behave as they do. Consequently, within development studies, this book, can be seen as contributing to the ongoing debate on the topic of comparative advantage in a number of ways (Cassen et al. 1986; DANIDA 1991; Fairman and Ross 1994; Jay and Michalopoulos 1989). Although I don't specifically compare the strengths and weakness of multilateral, bilateral, and nongovernmental organizations, I do raise a similar notion by using the sociological concept of organizational fields (DiMaggio and Powell 1983). Organizational fields may provide an even finer-grained separation of organizational types than those classified simply as multilateral, bilateral, and nongovernmental. I discuss this more in the final chapter.

Natural Resource Sociology

From natural resource sociology a number of influential works with organizational foci come to mind, including Anderson and Huber (1988), Burch (1971), Kaufman (1960), Poffenberger (1990), Selznick (1949), Schiff

(1962), and West (1982, 1994). The research by Burch, Selznick, and West represents the "old institutional school" as established by Selznick's classic study of the Tennessee Valley Authority, *TVA and the Grass Roots* (1949). In the old institutional school, the case study approach is essential. There is also an exposé, character to this type of work, because the premise of this school is that "things are not as they seem"; issues of political economy take centerstage (Perrow 1986, 159). Selznick showed that the TVA was willing to compromise its performance for purposes of survival by letting local constituencies control certain secondary natural resource management activities in return for their indispensable political support in maintaining the project's primary goals of flood control and energy production. Burch (1971) and West (1982, 1994) continued this line of investigation with comparative analyses of a number of U.S. natural resource agencies, the nature of their constituency relationships, and the particular consequences these relationships had on the social stratification in certain rural areas.

While these studies followed the theme of constituency power—and, therefore, institutional control—in modern democracies, other studies such as those by Schiff (1962) and Kaufman (1960) looked more closely at the internal functions of organizations and their effects on the organizations' resource conservation missions. In these studies, the characteristics of an organization's internal structure and professional procedures shaped its actions and performance.

Yet another line of investigation is concerned with how organizations can force their will on others. In his classic book *The Forest Ranger* (1960), Kaufman argued that it was in large part the professional training of the forester—or what I would call the profession's value-laden core technology—that allowed the very decentralized U.S. Forest Service to work as effectively as if it were a single, centralized organization. In short, the particulars of a given organization make a difference in the nature and level of the work it carries out while tackling a given task.

Political Science and International Organizations

For reasons unknown, international organizations, even well-known and influential ones involved in economic development, have not been studied much from a sociological perspective (Brechin 1989; Le Prestre 1985; Ness and Brechin 1988).[6] A number of political scientists, however, have long been interested in multilateral organizations, though they take a different path. For example, neorealists (e.g., Waltz 1979) see international organizations as insignificant or at best nothing more than the simple extensions of powerful states, existing simply to do the states' bidding. Such a potent but coarse-grained view is likely to gloss over the more refined subtleties that can play an important role in shaping international events. The criticism that the Bank has received from NGOs and others regarding its environmental effects, on the one hand, or FAO's active political representation of Third World perspectives, on the other, serve as good counterexamples.

Other political scientists have proposed a "functional theory" that states that the proliferation of "specified-function" international organizations will create a more peaceful and integrated world system (e.g., Hill 1978; Jacobson 1984; Luard 1966; Mitrany 1966). They argue that as more international organizations become involved with problems of economic inequities, there are greater opportunities for achieving economic justice and peaceful coexistence through greater integration, communication, and the mutually beneficial exchange of resources. Organizational sociologists, on the other hand, argue that these scholars need to view organizations more organically and less mechanically. International organizations should be viewed not simply as tools that quietly promote integration but also, possibly, as recalcitrant or even contrary forces working against change—or at least certain types of change (Ness and Brechin 1988). Organizations, as collectives, often use their resources to define and maintain their own self-interest as they struggle with internal and external forces for control of the organization, its resources, and its outputs (see, e.g., Perrow 1986; Selznick 1949, 1957; Scott 1987; Zald 1970). From this perspective, differences in capacities and levels of performance are to be expected among organizations. When this perspective is applied to international organizations, it suggests that not all of them will contribute equally to the creation of economic welfare and the process of world integration. Conceptually, at least, some organizations should be viewed equally as possible impediments.

Recent works on international regimes (see, e.g., Krasner 1983; Young 1989), international conflicts and cooperation (Kahn and Zald 1990), "epistemic communities" of scientific consensus (Haas 1992), organizational knowledge (Haas 1990), international organizations as "mesoinstitutions" in international politics (Abbott and Snidal 1995), and especially the works that evaluate the effective performance of international environmental arrangements (Haas, Levy, and Parson 1992; Levy, Keohane, and Clark 1994; Mitchell 1994), have moved the study of international organizations closer to a sociological perspective, but much more can and should be done.

Institutional theory, however, provides a fertile area for uniting sociology and political science around the notion of international organizations.

Although a number of disciplines seem to have their own distinct notions of institutional theory (see Campbell 1994; Dimaggio and Powell 1991; Hall and Taylor 1994), political science's contribution to policy paradigms and their shift (Hall 1993) has particular relevance to my study. Hall's work provides crucial insights into how shifts in policy paradigms occur. This sets the stage for a more sociological look at how organizations actually adopt new strategies configured to achieve new policy goals. This process is discussed in more detail in the final chapter.

A more applied approach, however, would be the need to understand the inner workings and sociological arrangements of international organizations so that we may better ascertain the potential of these organizations for effective governance of the global environment. The ascendancy of the UN in these first years of the post-Cold War era points to the increasing importance of international organizations as problem solvers and social actors, nowhere more so than in the realm of resource management in lesser-developed countries and in the realm of global environmental governance as a whole. A sociological review of how these organizations behave will be helpful in understanding their potentials and drawbacks.

Lack of Empirical Foundations

The theoretical literature on organizational sociology would then suggest that the carriers of development programs have a very important role in determining the type of activities pursued and the overall success of the projects. Yet it seems that organizational issues of development are still being seriously neglected by observers. As noted by one of the reviewers of the original manuscript of this book, when development efforts fall short of expectations, students of the profession tend "to round up and blame the usual suspects—shifting cultivation, peasant fatalism, corrupt local politicians, neocolonialism," and the like. This book attempts to add one more suspect to the list: the development agencies themselves. When observers have focused on development organizations, they have tended to evaluate mostly microlevel administrative issues related to development or the actual implementation of field projects, such as the specific inadequacies of a given project's monitoring protocol. Still other observers have focused attention on the institutional inadequacies of recipient agencies or host governments (Hage and Finsterbusch 1987). In my study, a more contextualized analysis of the development organizations themselves is undertaken by reviewing the broader influences that shape those organizations. Here the focus is not on determing whether the most adequate measures have been established for project monitoring or evaluation but rather on understanding the larger forces and structures that would encourage or discourage effective monitoring in the first place. Hence, my work attempts to add more empirical evidence, which is severely lacking at present, to try to understand the performance and behavior of complex development organizations.

Organizational Elements Defined

By examining some of their salient characteristics, I try to understand the work and behavior of the three organizations. In particular, three common elements—technology, structure, and environment—take centerstage in this analysis. Another less-central element is organizational character, though it is helpful in summarizing the unique nature of each organization.

Although the technical school of organizational analysis is somewhat out of favor at present, I argue that the effects of an organization's technology on behavior and performance, at least for some organizations, can be more powerful than is presently acknowledged. This seems especially so in helping to determine how an organization internalizes those changes required of it by the external environment (Brechin forthcoming). Directly, I attempt to explain more precisely why differences exist between the organizations, especially when each has rural development forestry programs, encouraged by the same forces found within the larger institutional environment.

In my analysis, I speak of an organization's *core technology.* A core technology consists of the tools, procedures, and skills assembled to transform the inputs to the organization into outputs, or the process of turning various "raw materials"—be they metals, ideas, or other human skills—into useful products or services. *Structure* represents the organization's formal division of labor—in this case, its degree of centralization and formalization. Where is the staff located—at headquarters or in the field? Are there formal bureaucratic procedures that must be followed to the letter, or does the staff have considerable flexibility and authority to carry out their assignments? And so forth. In addition, six notions of organizational *environment* are utilized. Three of the more critical ones are *task, technical,* and *institutional* environments. The other three have to do with level of analysis: *micro-, meso-,* and *macrolevels,* from smallest to largest.

The concept of organizational environment emerged with greater clarity in the 1970s with the revelation (among academics especially) that organizations do not exist in social vacuums; that organizations interact with other organizations, as well as with the larger cultural and political forces of society as an essential part of their existence; and that organizations are greatly influenced by their relationships with each other. The concept of *task environment* attempts to capture an organization's immediate external relationships with other forces, in particular other organizations that are associated with work on a particular activity, whether it is acquiring raw materials or distributing some defined output. Although prominent theorists (e.g., Scott 1992) suggest that *technical environments* and task environments are the same, to me they are not. Technical environments are the social construction (or reality) of organizations that exist in settings where outputs and inputs are very closely linked; that is, performance in the production of outputs more likely determines an organization's survival. Obviously, organizations within dynamic marketplace settings exemplify the technical environment. The technical environment, then, is conceptually distinct from the task environment, which for me represents the set of external organizations with which the focal organization must interact in order to remain viable. Therefor, the organizational relationships that the task environment represents can be found in both technical and institutional environments, discussed next. The *institutional environment* represents the larger and broader influences from society, such as cultural norms, paradigm shifts, or sweeping governmental policies. Within an institutional environmental setting, the widespread adoption of new ideas, practices, or approaches is more likely to be linked to cultural symbols or norms that enhance legitimacy or prestige over economic or performance efficiency.

Organizational character is a less precise element to define. It is the summation of the effects of the internal and external forces that have shaped any given organization over time. Selznick (1957) notes that an organization's character develops when it becomes "infused with value." Here the organization gains an identity, a sense of purpose, and becomes an end in itself. It is no longer simply a means to an end, a mechanical tool of production, but a "living" organic entity whose survival internal and external interests view as the only true goal. Organizational character, in sum, represents what the organization is about—its founding principles and the strategies that it promotes.

While these four factors (core technology, structure, environment, character) sum up the organizational attributes, the notion of an organization's *task* must be raised. In my study, the task is the "job" that the organization works on. Here, poverty alleviation through forestry projects is the task. The nature of the task clearly separates the task from the organization itself. This is a key distinction for my study as a whole, especially when I discuss versions of contingency theory in the last chapter. As I discuss later, each task has its own set of characteristics, which may blend in well with those of the organization and its outputs—or gnaw and grind like mismatched gears.

The Task: The Alleviation of Poverty through Rural Development Forestry

Traditionally, forestry and forest management have focused on producing trees for industry and related commercial purposes. Although these practices of traditional or industrial forestry remain essential for maintaining a continuous supply of useful wood products for modern society, such properties provide little for the daily needs of the billions of rural people throughout the world's poorest countries and regions. These people make very different demands on forestry. To help meet those demands, rural development and community forestry emerged in the late 1960s and early 1970s.

Community forestry, also known as *social forestry,*[7] focuses on producing and managing trees to satisfy the basic needs of people in poor rural communities. It is, quite simply, "trees for people" (Westoby 1975, 1987). It is trees of, for, and by the people. *Rural development forestry* is meant to capture a somewhat broader notion than community forestry, but neither term overlaps with traditional industrial forestry. Both terms denote programs that attempt to address the needs of rural areas, rather than those of cities or corporations. In both, the funding for the program may come from an outside source, but the benefits are intended to accrue to local rural peoples. As used here, the difference is that community forestry programs are, by definition, *controlled directly* by the local people, whereas rural development forestry programs do not necessarily have to be. Examples of rural development and community forestry include government- or community-owned fuelwood woodlots or plantations, windbreaks or dune stabilization projects to control desertification in more arid regions, and the intercropping of trees with agricultural production.

Community and rural development forestry represent a revolutionary change in the forestry profession, requiring the development of innovative approaches (Bentley 1993; Fortmann 1986; Leach and Mearns 1988; Shepherd 1992). Because of the complex social interactions with local peoples, rural development forestry generally and community forestry specifically are seen by many as most successfully promoted through flexible "bottom-up" development efforts that require considerable sensitivity and responsiveness to fluid local conditions (see Brechin 1989; Leach and Mearns 1988). In sociological terms, community forestry is a complex and unpredictable task, found in a turbulent microlevel environment, consequently requiring complex, nonroutine responses. Organizations need to be well matched to the task if they are to perform well. Contingency theory predicts successful performance when a nonroutine, complex task like community forestry is matched with an organization that has a decentralized structure, allowing for more professional expertise and decision-making authority close to the action. As I discuss in Chapter 6, these qualities clearly distinguish the organizational outputs (community forestry projects) from the organizational task (poverty alleviation) and from the organization's core technology (professional knowledge). Nonetheless, contingency theory, too, only provides at best a partial explanation, especially when we review the efforts of FAO Forestry. Once again, the fullest explanations of organizational behavior come when we unite a number of theoretical perspectives.

Chapter 1

Using Trees to Move Money: The World Bank

The World Bank is a massive and complex organization. I focus only on some of its work in forestry. With forestry making up a minuscule portion of the Bank's loans, it can hardly be claimed that this sector has wielded much influence in molding the Bank's structure or shaping its behavior in any significant way. Rather, just the opposite is true, which is one of the main points I want to make. I argue that the character of the Bank has prejudiced what it can do in forestry. In particular, the Bank's core technology, and its investment approach to development in particular, has significantly influenced its work and performance in community and rural development forestry in the 1970s and 1980s. This is not to say that its structure and external relations are unimportant, for they, too, play important roles. The Bank's technology, however, seems to have played an exceptionally strong part, particularly when compared with that of the other two organizations; and given the lack of stature that technical perspectives have in organizational analysis today, this is a rather bold statement. Still, to understand the Bank's character and its involvement in this sector, one needs to look closely, although selectively, at its history and operating procedures.

But first let us reflect upon the World Bank's work in rural development forestry—in particular, how it became involved in it.

Community and Rural Development Forestry within the World Bank

The Bank has a long history of supporting forestry-related projects in lesser-developed countries (LDC). Its very first loan to an LDC was in 1953 for a pulp and paper mill in Chile (Brechin 1989; see also World Bank 1991b).

This, however, was more an industrial project than a freestanding forestry one. That kind of project first came about fifteen years later in fiscal year 1969 when the Bank provided a $5.3 million loan to Zambia for an industrial plantation. Only in 1974 did the Bank finance its first rural development forestry project, with a $2 million loan to the Philippines for small-farm production of wood to feed a local pulp mill.

The Emergence of Rural Development Forestry

The emergence of rural development forestry at the Bank can be traced directly to its president at the time, Robert McNamara, especially to his speech at the Bank's Board of Governors meeting, at Nairobi, Kenya, in the fall of 1973. (Of course, the ultimate roots of this shift in policy lay elsewhere in broader cultural shifts, discussed below.) In this famous speech, McNamara presented his vision of the Bank's new mission in attacking rural poverty. Poverty alleviation became a major focus within McNamara's Bank. His concern for poverty, inequitable development, and rising population had already been designated as important issues to be addressed during the early part of his presidency. His plan for addressing rural poverty through development more or less spearheaded the "second half" of his tenure (1974–81). In one sense, poverty alleviation had been a long-standing goal of the Bank, if one views economic development through industrialization and modernization as an attempt to eliminate poverty. McNamara's approach, however, represented a significant new direction to that effort. Instead of focusing on what had been the traditional mainstay of development, industrialization followed by "trickle down" development, McNamara's new approach would address poverty head-on through activities geared to the "poorest of the poor." Although there have been frequent debates on the effectiveness of the Bank's "poorest of the poor" strategy, some observers have argued that the Bank's lending program during the 1970s went through, at a minimum, a qualitative shift in emphasis (see Annis 1986; Ayres 1983; van de Laar 1980; Yudelman 1985).

Although some innovative twists in the Bank's lending activities did take place in agriculture before the McNamara era, a large portion of its activities remained focused on developing the industrial capacities of cities. Most of this activity centered on the construction of basic economic infrastructure, such as roads, dams, ports, communication facilities, and factories for heavy industry. During this time, the rural areas were viewed chiefly as sources of cheap labor that would help fuel the drive to industrialize; in fact, rural areas were intentionally ignored (Lewis 1954). Capital was not considered a limiting factor to agricultural production. It was expected that as the cities and their urban markets grew, the rural areas would benefit from an increased demand for agricultural products and a general increase in economic welfare (see Yudelman 1985, 2). In short, the Bank, and other development aid organizations, did little in agricultural development until the 1960s, and most notably the 1970s.[1]

Process of Institutionalization: The World Bank Shifts to a Rural Poverty Paradigm

With the publication of several influential reports and studies, including the Pearson Commission Report, released in 1969, experts began dealing devastating blows to the established industrialization model of economic development (see also Berg and Gordon 1989). These reports pointed out both the inadequacies of the industrial development approach and the inequalities of its effects. The most serious failing of the Bank's old lending strategy, which closely pushed and reflected these established theories of development, was its creation of urban islands of relative wealth, surrounded by seas of rural poverty. Agricultural production was lagging behind rapidly rising populations, creating the specter of serious food shortages, which in the 1960s became known as the "world food crisis." In addition, industrial development proved to be a slow and uneven process, leaving the large and rapidly growing rural labor markets underutilized.

The lot of the rural people in the developing world was deteriorating rather than improving. To generalize, successful growth in the Gross Domestic Product was often paralleled by poor local distribution. At the very least, growing rural poverty was negating any gains experienced elsewhere in the economy. It eventually became increasingly obvious to theorists and practitioners that the future of economic development of many developing countries was to depend, at least in part, on the health and well-being of the rural sector and on the production of food. From this realization emerged the rural development perspective within economic development thought. At the same time, after experiencing the tragedy of food shortages, many developing countries desired to be more self-sufficient in food production and demanded greater international assistance in agriculture. It was these events that led to McNamara's Nairobi speech in 1973. From that point, rural poverty and its elimination through directed rural and agricultural development programs became central themes in the Bank's lending policies.

When McNamara hailed the new agenda for poverty alleviation, he provided, as one senior Bank official put it, a "new score" for Bank performance, one which sent the staff scurrying to come up with development projects that addressed the issue of rural poverty. The agricultural sector became the centerpiece of McNamara's program to combat rural inequities. In the early 1960s, the Bank's programs in agriculture made up only 6 percent of its total yearly lending activities.[2] By the mid-1970s that figure rose to 30 percent, making it the largest single component in the Bank's lending portfolio. Under McNamara, loans to agricultural programs increased from $400 million a year in 1970 to $3 billion a year in 1980 (Ayres 1983; Yudelman 1985). Raising the productivity and income of the small farmer became important parts of the Bank's effort (see Yudelman 1985).

Actually, McNamara's movement into rural development was more the culmination of an evolution in development theories that began in the 1950s (and which had been given credence by former Bank presidents as well as others) than a truly revolutionary change. The turn to rural poverty, nevertheless, represented a bold new step in the Bank's programming. McNamara's strategy, however, is as much a response to the failure of the Bank's previous strategies as an attempt to mitigate the unanticipated consequences of its previous policies. Perhaps most important, it was a way to develop new "markets" for the Bank by expanding its lending portfolio.

Forestry, Poverty Alleviation, and the Environmental Revolution

Not surprisingly then, the Bank's movement into rural development forestry was a natural consequence of its shift to poverty alleviation as an appropriate response to new theories and polices in economic development. Yet, rural development forestry had its own unique set of driving forces. Its rise was linked to the confluence of several crucial events. Chief among these were the energy crisis of the early 1970s; the United Nations Conference on the Human Environment, held in Stockholm, Sweden, in 1972; and a number of the publications on environmental problems in Third World countries. These events, as well as the desire to assist small farmers, were elements of McNamara's framework.

When the Organization of Petroleum Exporting Countries (OPEC) precipitated the energy crisis of the 1970s, shock waves were sent around the world. Fuel prices went up not only for the developed countries but also for many developing countries. Many oil-deficient Third World countries were forced to find ways to bolster a balance of payments that had been sent reeling by the increased price of imported oil. The basic economic strategies of the time were to (1) increase exports, (2) reduce imports, (3) borrow, or (4) employ some combination of the three.

Forestry was seen as a partial solution, but in a somewhat different light (see World Bank 1978). With rising fuel bills from imported oil, pressures mounted within poor countries to find and use more domestic sources of energy. Fuelwood was an obvious substitute. The use of wood as a fuel had been recently recognized as a major source of energy in the developing world, particularly in rural areas. The appreciation of wood as a source of energy had been growing since the 1972 Stockholm Conference. Its importance was highlighted by Erik Eckholm in his 1976 landmark publication entitled *The Other Energy Crisis* (see Arnold 1992; Leach and Mearns 1988). Eckholm not only reemphasized the vast dependence of millions of rural and urban poor on wood as their chief source of fuel, but noted its increasing scarcity and some of the social, economic, and environmental consequences.

The scarcity of fuelwood was forcing the use of crop residues and animal dung for fuel instead of as fertilizer, important in maintaining soil fertility. The demand for fuelwood began to outstrip supply. Rural people had always depended upon the natural regeneration of forests and woodlands or upon newfound resources for their supply of fuelwood and other wood products. But, forests, ironically, were often being destroyed to create more agricultural land. This situation combined with the growing number of people to create the fuelwood crisis, which continues today.

The Bank's interest in rural development forestry was not so much an interest in the environment or environmental protection as it was a concern about the economics of oil and strategies to overcome the need for it. This fact had a profound effect in shaping the Bank's rural development forestry program, especially in its early years, when it focused on establishing fuelwood plantations (Arnold 1992; World Bank 1978, 1991b). At the same time, the important link of forestry to agricultural production was not completely ignored by Bank officials. Since the trees were needed most in the rural areas, and given that most of the land was over cultivated or otherwise used by rural people, it became essential for rural people to be meaningfully involved in the planting and management of trees. This involvement fit nicely into McNamara's program of helping small farmers—at least as it was originally planned. As we shall see, the reality of the Bank's effort paints a different picture.

In 1978, the Bank produced its first forestry policy sector paper, which provided the political justification for its lending in forestry and promoted the need for rural development forestry over more traditional forestry programs (World Bank 1991a, 1991b). The sector paper provided greater coherence to an otherwise sporadic program. As a result, forestry generally, and rural development forestry particularly, became sanctioned activities, though still a minuscule part of the Bank's development efforts.

In sum, the Bank's movement into rural development forestry was the result of McNamara's directives. It was a top-down action, emanating from a large and powerful centralized organizational structure. Although macrolevel events such as the energy crisis, the onset of fuelwood scarcity, changes in rural economic development theories, and even FAO's involvement in forestry provided ultimate sources of change behind the Bank's push, McNamara and his staff were the proximate source.

As noted in the first chapter, this top-down movement into community forestry is in contrast to the more decentralized, "bottom-up" movement exhibited by CARE USA, and the combination of the two as seen in FAO Forestry. At a more theoretical level, the macrolevel processes of institutionalization were behind the change for each of the three organizations. But to emphasize again, the particular manifestations that this change had on the organizations' behavior and performance was unique given the specific internal elements and task environment elements found in each organization. These paradigm shifts in economic development theory and environmental concerns influenced not only the Bank's activities but provided the larger cultural setting for the two other organizations.

The Character of the World Bank: A Brief History

The World Bank, formally known as the International Bank for Reconstruction and Development (IBRD), was created at a meeting of 44 nations (the Allied Powers and their supporters) at Bretton Woods, New Hampshire, in July 1944. In 1946, the Bank opened its doors in Washington, D.C., where it has remained along with many of the other international banks and financial organizations.[3] It is one of four organizations that constitute the "World Bank Group." Another is the International Development Association (IDA), which is often referred to as the Bank's "soft-loan" affiliate. The IDA was created in 1960 to provide loans (or credits, as they are called) at concessional terms to the poorest nations. The other two members of the World Bank Group are the International Finance Corporation and the Multilateral Guarantee Agency (MIGA), which were created in 1956 and 1988, respectively. These two agencies operate under a different set of criteria and management personnel. Our discussion will focus only on the activities of the Bank and IDA.[4]

The Bretton Woods meeting in 1944 capped some three years of preliminary work and negotiation on issues of international finance (Mason and Asher 1973, 1). Its goal was to formalize an international banking scheme. It was only one of a number of meetings on the establishment of the world polity that became known as the United Nations.

The meeting members, greatly influenced by the desires of the U.S. delegation, hammered out final agreements on the creation of international organizations to provide greater stability to international financing activities and to finance the reconstruction of Europe and the development of the less-developed countries. Interestingly, the main agenda item for the meeting was the establishment of the International Monetary Fund (IMF), not the Bank. Discussions on an international bank were to be pursued only if there were sufficient time to do so (see Mason and Asher 1973, esp. chap. 2).[5]

To understand the World Bank and its operations, one must view it first as a bank. As Ayres (1983) points out, "The World Bank has always been something more and something less than a 'real' bank" (10). And to fully appreciate this statement, one must understand the times of its creation, especially serious concerns held by the international banking community.

The International Investments Crisis

A dark cloud hung over the world of international finance during the 1940s. Prior to the interruption by World War II, international banking in the 1920s and 1930s was plagued by gross mismanagement, questionable activities, wastefulness, and defaults (Mason and Asher 1973). In particular, all confidence in international investment was lost with the Great Depression of the 1930s. The IMF and World Bank were viewed as the principal vehicles for restoring confidence in international investments in the postwar world. The IMF was to bail out countries suffering from temporary balance of payments deficits through the availability of short-term loans and economic policy adjustments. The Bank was to take a more long-range view. It was to be a financier of last resort, with the mission of rebuilding destroyed economies from the war and of nurturing fledgling ones as new countries were born.

It was hoped that through sound lending procedures and banking competence the IMF and World Bank would restore the lost confidence in international investment. For the Bank, the concept of *project loans* was viewed as the innovation to help accomplish such a task. Loaning capital for specific, stated projects was considered an important means of controlling the quality and accountability of the loans (Baum and Tolbert 1985; Mason and Asher 1973). The desire to move money through project loans, with particular interest in sound economic principles, became woven heavily into the fabric of the Bank.

The World Bank as a Bank

The World Bank today is a multifaceted organization. It is both a bank and a development agency. Officials at the Bank have neatly coined its raison d'être as "investing in development" (see Baum and Tolbert 1985). Although the Bank's role as a development agency has grown through the years, most observers and Bank officials themselves agree that the Bank is a bank first and development agency second. This was true especially in the pre-McNamara days (before 1969). Even today, the Bank's major concerns are with issues of interest rates, rates of return on investments, repayment schedules, creditworthiness of borrowers, and the protection of its investors. As Ayres (1983) has keenly pointed out, the Bank "does not give things away" (10).

It is also something less than a bank in the conventional sense. The Bank does not accept deposits nor does it make loans to individuals. Sovereign nations become its members by purchasing voting shares. Loans are given only for purposes of economic development and only to developing countries that meet certain requirements. Yet, most of its operations are essentially financial in nature, and sound economic analysis is its hallmark.

The Bank's preoccupation with financial and economic concerns is not by accident. As detailed by Mason and Asher (1973), the Bank was established as an institution that had to borrow in order to lend. Consequently, it is a conservative institution. Since it depends almost exclusively on the world's financial markets for investors in its securities, it worries about its credibility as a sound financial institution. In the first ten years of its existence, it depended almost exclusively on the U.S. financial market, a principal reason that the Bank is located in the United States. Today, however, the Bank operates within all of the major financial centers and even strikes investment deals with wealthy individual governments.

One of the central consequences of its banklike character is the Bank's need to move money. The Bank exists financially, in part, by turning over loans. To repay its creditors, the Bank must lend money and collect both principal and interest from its borrowers. It also sells its loans to private investors in the securities markets. Thus, getting its loan portfolio out as quickly and as efficiently as possible has been a major Bank objective that colors its internal operating procedures. Consequently, the need to move money has tended to bias the Bank's lending activities to larger loan projects and to those that require less staff time to prepare and supervise.[6] As we shall note later, this bias has affected the Bank's forestry lending program just as it has other sectors.

The World Bank as a Development Agency

Ayres (1983) argues that the World Bank has always been something more than a bank. Very early in its existence, the Bank was forced not only to finance projects but to become actively involved in their identification and preparation. Its role in development-related technical assistance has expanded throughout the years to become an extensive and integral part of the Bank's operations and character (see Mason and Asher 1973, esp. Chap. 10).

In 1960, at the urging of the United Nations Economic and Social Council, the Bank created a "soft loan" affiliate, the IDA (Mason and Asher 1973, 97). Its purpose was to provide development loans to the poorest of the poor countries. These countries would not normally qualify for more conventional commercial or even Bank IBRD loans. The IDA loans, or credits as they are called, are at concessional rates. In addition, the sources of the credits differ from IBRD's loans.[7] The creation of the IDA was an important signal that the Bank was becoming a development agency (Mason and Asher 1973, 87). These differences have an important influence on certain loan programs since IDA credits are more difficult to obtain and the size of the fund is not as large as IBRD's.

The Bank's growth, especially under McNamara (1969–81), pushed the Bank into an even larger array of development activities. McNamara steered the Bank away from the narrower development concept of strict capital accumulation through the construction of economic infrastructure to greater investment in a wider range of development-related activities, including human capital. Rural development forestry obviously falls within this broader view. The movement into less-traditional areas troubled some who saw the Bank straying too far away from capitalistic notions of economic development. Some even labeled McNamara's efforts the "International Great Society" (see Phaup 1984, 13). Yet, others have argued that the

Bank was trying to reflect a more complex understanding of the process of economic development because of its lack of success after years of following the narrower path (Ayres 1983, 233).

Regardless of one's feelings about the context of the Bank's expanded program, its growth has made the Bank the world's leading authority on the process of economic development and on the economic activities of developing countries. Based upon its own research, observations, and experiences, the Bank issues influential sector policy papers. These policy papers, and the annual addresses by the Bank's presidents, have for decades set the tone and path of development activities followed by others. In addition, the Bank collects and summarizes valuable information that is important for understanding the conditions of economic development throughout the developing world. This information is made available through published annual reports, such as the Bank's *World Development Report* and its prestigious country economic reports. In addition, because of its financial resources, the Bank has the ability to operationalize its own policy decisions (Ayres 1983, 10). This has allowed the Bank to play a leadership role in matters of economic development. Overall, these activities and capabilities have helped make the Bank the world's premier economic development organization, in addition to its being a leading international financial organization. Its movement into the development arena has opened it up to criticism from conservatives who fear it has lost or will shortly lose its capitalistic approach. Likewise, the Bank has always faced general criticism from liberals who have long wanted the Bank to do more in the name of development (see Ayres 1983; Hayter 1971; Phaup 1984). The differences between these two perspectives have caused some debate inside as well as outside the Bank (see World Bank 1992b, "Wapenhans Report"). The tension has become an important characteristic of the Bank. It continually struggles over its need, on the one hand, to be financially sound, to appeal to the conservative nature of the financial world, and to move money for personal promotions, while at the same time it desires new programs to assist member countries with their wide-ranging development needs. This tension over its mission has had important ramifications for its lending programs, including those in forestry.

The World Bank's Staff

Although both conservatives and liberals criticize the World Bank for what it does or does not do, most would acknowledge the professionalism of its staff. Mason and Asher (1973) state: "Over the years, the Bank has acquired a staff unique among the international agencies in terms of its professional competence" (71). The Bank, in general, has always maintained a staff with high standards and considerable technical competence. In particular, the Bank is unique in that it is home for more superb economists than perhaps any other international organization (Mason and Asher 1973, 71).

In the Bank's early days there were only a few economists. The bulk of the Bank's project staff in the 1950s, for example, consisted of engineers, whose presence made some sense given the Bank's preference for technical, infrastructure projects at that time. The economists then on staff were located in the area departments that were working on country economic forecasts (Mason and Asher 1973; van de Laar 1980). From the very beginning, however, the Bank has realized the need to acquire a qualified technical staff. Baum and Tolbert (1985) state that in order to retain the confidence of the banking community, which purchases the Bank's securities, it has been essential for the Bank to assure that the projects are technically and financially sound (9). Other experts point more to the internal culture and professionalism of the Bank staff as the force striving for technical and financial competency. The international banking community is more likely to be concerned about the general financial health of the borrowing country and not the particular analysis of any given project. Still, with the increase in the size and scope of its lending over time, especially during the 1960s and 1970s, the Bank had to increase dramatically the size and composition of its staff to maintain its technical competence.

Over the last several decades, the Bank has ventured into new programmatic areas like agriculture, education, population and health, and forestry. Professionals were hired to reflect the new work in these areas. In the early 1980s, specialists representing over 80 different disciplines were employed, with two-thirds of the professional staff engaged in project work (Baum and Tolbert 1985, 9). But it was the rise of more sophisticated economic project appraisal methods in the late 1950s and early 1960s that caused the Bank to hire more professionals with economic backgrounds. Their skills in the economic appraisal of projects became more crucial as the Bank began to rely more heavily on that type of analysis.[8]

Compared to the international representation of other United Nations organizations, the Bank is still grossly tilted in its professional positions toward the Western industrialized countries (van de Laar 1980, 99; see also World Bank annual reports). Using 1994 records from the personnel department, 60 percent of the Bank's 2,455 professional staff were from Western countries. Some feel that this characteristic makes the Bank more a "Western Bank," with an air of arrogance and paternalism in its presumption of how best to establish economic order among the developing countries (van de Laar 1980, 103). Still others remark that greater internationalization of the Bank's staff would have little effect, since such paternalistic behavior is structural in form, given the basic nature of the Bank's work and its business relationship with its clients. Although both arguments may be at least partially true, others convincingly argue a more political reality, namely, that greater internalization of the Bank's staff may seriously threaten its chief supporters: Western countries. It has been suggested by some that a movement in that direction would result in the withdrawal of Western countries' financial support for the Bank, prompting those countries to push alternative funding organizations or to make greater use of bilateral assistance (van de Laar 1980, 104). As we shall discuss later, the Western countries, particularly the United States, maintain a fairly tight political control over the Bank. For example, all Bank presidents thus far have been American. In the end, the Bank needs to maintain the confidence of the Western financial markets, where it raises a lot of its working capital.

In summary, the Bank is a Western organization where top-flight project economists have reigned supreme since the 1960s. The professionalism of this group galvanized the Bank's operations in framing project loan agreements under rather strict benefit-cost terms that often require inflexible project designs. Project economics at the Bank began to wane a bit in the mid to late 1980s with the rise in structural adjustment, sector lending, and criticisms of the impacts of certain Bank-sponsored projects. Although this criticism of the Bank's social and environmental consequences is forcing the Bank to adjust its approaches, the project approach still carries considerable influence within the Bank's professional culture and, specifically, in its work in the forestry sector.

The World Bank's Structure

According to a former Bank official, Aart van de Laar (1980), the most characteristic feature of the Bank is its centralization (215). In spite of its 52 resident field missions and three regional field offices, two in Africa and one in Asia (see World Bank 1992a), the Bank has maintained the vast majority of its staff at its Washington, D.C., headquarters. Ayres (1983) estimates that in the early 1980s 94 percent of the professional staff was located in Washington (64). It is still the same today. Most of the Bank's centrally located staff work on the development of project loans (Baum and Tolbert 1985, 9). It is a massive organization with a formalized bureaucratic structure that operates in a top-down manner (Ayres 1983, 65; Rich 1994).

The nature of the Bank's structure has affected its ability to conduct certain types of development activities—especially its work in poverty alleviation (Ayres 1983; Clements 1993; van de Laar 1980). Because of its dramatic growth in the 1960s and early 1970s, the Bank underwent a serious reorganization in 1972 to improve its internal operations.[9] The reorganization essentially placed the project personnel, who had been assembled in sectoral units according to their expertise, into area (geographic) departments. Among other things, this reorganization *decentralized* the Bank's work by geographic region, but in a purely *centralized manner.* The reorganization took place at the offices in Washington and did not reallocate staff to field units (see Van de Laar 1980, 215–222). It was done to improve the Bank's internal operations by facilitating the movement of its money. It was not undertaken to improve the design of projects to reach the poor (King 1974, 34; Van de Laar 1980).

The centralized nature of the Bank tends to hurt its work in poverty alleviation, including its work in rural development forestry. Success in these types of projects requires greater local-level knowledge and expertise, which "the Bank arguably did not possess in sufficient abundance" (Ayres 1983, 65). The design and implementation of poverty projects also require greater participation by the recipients. (Ayres 1983; Clements 1993). The Bank's lack of a decentralized structure has constrained its potential for incorporating grassroots elements into its projects. As a result, many of the Bank's poverty-oriented projects, including those in rural development forestry, have been seriously flawed.

In summary, the Bank is a highly centralized institution with its work decentralized by geographic region. The majority of its staff work out of the Washington, D.C., headquarters where they concentrate on moving money through project loans. These important structural features, the characteristics of its task environment (discussed next), and its formalistic technology play pivotal roles in determining the qualities of its activities.

The World Bank's Task Environment

Organizational sociology stresses that no organization is able to isolate itself completely from others and thereby eliminate the potential for dependencies. The World Bank, relatively speaking, is a self-sufficient and powerful organization. Mason and Asher (1973) remark that the Bank often dominates its relationships with other organizations (539). As already noted, the Bank has the ability to generate its own financial resources through investments and through the "turning over" of loans made to developing nations. In addition, it has a formidable reputation for negotiating with nations that are often desperate for cash (to keep the machinery of government and the economy running), as well as for promoting long-term economic development. These realities give the Bank considerable leverage in the management of its affairs. But the Bank is not all-powerful. It does have a few major weaknesses that allow others to influence its activities. The two most important sets of organizations in the Bank's task environment are other international development organizations and the Bank's member governments.

Member Governments

To be a member of the World Bank, governments must first join the IMF. Once members of the IMF, governments join the Bank by purchasing voting shares called "capital subscriptions." The amount of capital subscription to be purchased is negotiated between the applicant and the Bank's Board of Governors (Ayres 1983, 58). Upon joining, the government pays in 10 percent of its subscription (1% in gold or U.S. dollars and 9% in its own currency). The remaining 90 percent of its subscription is callable, that is, subject to the Bank's appropriation if it needs it to meet its obligations (Ayres 1983, 58; World Bank annual reports).

So the voting power of individual members of the Bank is determined by the amount of the members' capital subscriptions. Each member receives 250 votes plus one additional vote for each share of stock it holds. One share of Bank stock costs $100,000 (in 1944 dollars) (Ayres 1983, 59). Member governments can increase their voting power by renegotiating with the Bank the purchase of more shares of voting stock. With this type of voting arrangement, Bank member governments can be placed in two basic categories: financial supporters and loan recipients.

The financial supporters who are, in effect, the controlling member governments, have traditionally been Western industrialized countries. As noted earlier, the U.S. government has always been the biggest investor in the Bank. This, of course, leads to an important point in understanding a serious limitation of the Bank. Although they possess little direct control over the Bank's daily activities, the Western governments, and in particular the United States, maintain considerable influence over the Bank's fundamental well-being, especially its continued growth (Phaup 1984; Rich

1994). To increase its financial endowment, the Bank, upon the consent of its Board of Governors, must seek formal approval from the member nations if it wishes to increase its capital subscription rates. The United States and other wealthy countries are obviously central to that effort. Raising subscription rates allows the Bank to generate more financial resources to increase lending. Consequently, the Bank is sensitive to criticism from its largest shareholders.

Since the early 1980s, members of the U.S. Congress have shown considerable interest in the Bank's performance owing to pressure from two groups. One group consists of the conservatives, who historically have not been enthusiastic about multilateral activities and who became increasingly suspicious of and dissatisfied with the Bank's activities during the 1970s and 1980s (see Ayres 1983; Phaup 1984). The other group is composed of environmentalists, who have documented the Bank's disregard for the environmental impacts of its loan programs. Since the 1980s the U.S. Congress has been influenced by both groups.

Responding to changes in international financial markets resulting from the debt burden of developing nations, on the one hand, and to conservative calls for reductions, on the other, the Bank again underwent another massive reorganization in 1987 in an attempt to decrease its payroll and improve its ability to facilitate structural adjustments lending. During the time that this reorganization was taking place, the U.S. Congress and environmental organizations criticized the Bank for its alleged lack of sensitivity to the environment (Brechin, forthcoming; Rich 1986, 1994). The Bank responded to these criticisms, establishing its Environment Department in 1987, was at a time when it was trying to substantially increase its subscription rates to raise more capital so it could increase its work in structural adjustments.

The loan-recipient governments also influence the Bank's operations. These recipients are the LDCs, which the Bank was created to assist. It is easy to view these nations as being very dependent on the Bank for needed cash and as lacking the political influence to shape the Bank's affairs. That is true. But the Bank also depends on the recipient nations to be vehicles for moving money. There is a great deal of variance in this power-dependency relationship, however. The actual nature of the relationship between the Bank and a client is contingent largely on the number and size of loans that a country can acquire. There has tended to be a positive correlation between recipient control over the nature of the Bank's loans and the number and size of loans made. Because the Bank needs to move its money, it depends heavily upon its favorite customers, especially India and Brazil (see also the discussion of the Morris Commission findings in Rich 1994). India's status as a large borrower has greatly influenced the Bank's work in rural development forestry. In addition, the recipient governments maintain some control over the type of loans to be issued, as I stress later.

By its charter (Article III, Section 2), the Bank is required to deal only with the recipient government's treasury or ministry of finance. It is with these officials that the Bank staff hammer out project activities and negotiate the specific loan agreements. Consequently, it is up to the staffs of the ministry of finance and the various implementing agencies to determine what are appropriate projects for their country and what are not. There may be differing perspectives among government ministries on what a country's development needs are and what should drive project loan selections. For example, Ayres (1983) discusses at length the problems that the McNamara Bank had in pushing its poverty-oriented program. Many government officials were simply not interested in helping the rural poor. Other officials failed to make the link between the elimination of poverty, on the one hand, and economic development, on the other. Crane and Finkle (1981) noted a similar reluctance during the 1970s on the part of certain country officials in the area of population planning. The Bank's lending program was influenced by the willingness of client countries to accept certain types of development activities over others. This same problem has emerged with forestry-related projects as well. A recipient government's need for generating financial capital, rather than other considerations, helps to drive the government's project-selection process. Consequently, the Bank faces difficulties whenever it attempts to promote certain loan programs that are not in demand by client governments (Ayres 1983; Crane and Finkle 1981).[10]

The nature of the Bank's relationship with its member countries also provides an important perspective on the criticism received from the Bank from the Right as well as the Left. Ayres (1983) notes: "Those who see aid as imperialism apparently assume that aid agencies can obligate recipient countries to do or not to do just about anything. Those who fault the Bank for not doing enough—in the amount of its lending, the kinds of projects it funds, or the limited amount of influence it sometimes seems to have in orienting the development policies of many countries—seem to assume exactly the same thing" (58). Clearly the Bank's relationships with its member countries affects the Bank's activities and overall performance. Unless Bank staff strongly object, recipient government officials can select or demand projects of certain requirements or can select projects simply for their financial value with little regard for their technical or social aspects.[11] This should not be particularly surprising since the basic relationship between the Bank and the finance ministries is to move money for their mutual benefit. This relationship remains very much intact. The question is less Will money be moved? than What form will it take?

Recipient governments influence the Bank's activities and performance in other ways as well. There exists a great deal of variance between a recipient country's ability to administer project loans and its ability to absorb them. This last point generally refers to the ability of a country to efficiently spend the loan through the agency's activities. Both these points have important implications for the Bank's performance in certain program sectors, including forestry. Forestry institutions in LDCs are for the most part very weak, and, consequently, developing countries have limited capacity to undertake Bank loans. This has severely limited the Bank's work in forestry, especially for rural development forestry (see Braatz 1985; World Bank 1991b). Without the structural vehicles to service large loans, Bank and ministry officials must find more conducive channels for moving money.

Other Development Organizations

Without question, the World Bank has become the giant of the world's development organizations. In spite of its size and dominance, the Bank and its activities have been influenced by other development organizations. Generally, the relationships have been ones of mutual benefit and cooperation, although Mason and Asher (1973) remark that early in its existence, it had to "slug it out at times with the Export-Import Bank of the U.S." as it attempted to develop clients overseas (535). Mason and Asher (1973) also state that the Bank has been helped by other development organizations, especially bilateral ones, that have been willing to finance the imports needed to make its projects viable (232). The Bank has even maintained a friendly and complementary relationship with its "Bretton Twin," namely, the IMF (see Mason and Asher 1973, esp. chap. 16). Finally, one of the more dramatic effects on the Bank was the creation of its "soft loan" affiliate, IDA, in 1960. The concept of IDA was pushed strongly by the United Nations Economic and Social Council (Mason and Asher 1973, 97).

Perhaps most important, the Bank continues to depend on other organizations as sources of fundable projects. It is important to note that a serious constraint in the Bank's development work has been, not lack of money, but lack of fundable projects to move money (Mason and Asher 1973, 73, van de Laar 1980, 235). The Bank is required by its Articles of Agreement to transfer at least 90 percent of its capital to LDCs through projects. Given the Bank's need to assure its financial credibility, the loan process is usually rigorous. Project loan proposals frequently have gestation periods of several years (Ayres 1983; Baum and Tolbert 1985).

In an attempt to overcome the project "bottleneck" problem, the Bank, very early in its existence, began developing partnerships with other development organizations. The Bank has worked very closely with UNDP, which finances the Bank's preinvestment country studies. These studies attempt to identify in member countries general opportunities for financial investments. The Bank has also entered into cooperative arrangements with the United Nations Educational, Scientific, and Cultural Organization (UNESCO), World Health Organization (WHO), and FAO to develop projects in agriculture, education, water supply, sanitation, and drainage (Mason and Asher 1973, 82). Many of the Bank's traditional forestry projects originate from the IBRD/FAO cooperative program. Since the mid-1980s, the Bank has tried to work closer with NGOs in an effort to pursue more grassroots development projects. Nelson (1991), however, has reported that the Bank was having difficulty in sustaining those relationships.

In summary, the Bank is a powerful organization and often dominates others in its task environment. Yet it relies on some organizations for crucial services. In particular, it depends on wealthy industrialized countries, especially the United States, to financially support the expansion of its lending operations and for replenishing its IDA funds. The Bank's operations can attract criticism from wealthy, contributing governments, thus affecting its growth, and can cause these governments to meddle in its affairs, although serious changes would likely be avoided given the Bank's strategic position in the international financial world. The Bank also depends upon the poorer nations as customers for investment. Because of great economic and financial need, it is unlikely that these countries will ignore the Bank and its resources. Still, these governments exercise considerable authority in selecting the nature of their indebtedness. This has weighty consequences on the type of loans the Bank makes and on the characteristics of its lending.

The World Bank's Core Technology: The Project Approach

The Bank's core technology can be described, in its barest form, as investing in economic development. This technology has been packaged more concretely by means of specific project-related investments, administered through a process called the "project approach" (Baum and Tolbert 1985; see also Crane and Finkle 1981).

The Bank moves its money through appraised development project loans. The Bank considers a project sound when the expected total economic benefits derived from the project exceed the expected total costs and when the project is arranged in a manner fiscally prudent for both lender and borrower. As noted by Ayres (1983), the Bank's project activities are the heart of its operations. Although the Bank has historically worked through project loans, the international debt crisis, worsening since the late 1970s, has forced the Bank to take a more active role in nonproject lending. A prime example is "structural adjustment lending," which is essentially an attempt to help keep countries afloat by providing medium-term loans for adopting economic policy adjustment packages geared toward reducing their current balance of payments at the macrolevel (for more details, see Ayres 1983, 42; see also World Bank annual reports). Another form of nonproject lending is "sector loans," also known as "program loans." They provide money to develop the borrower's capacity to conduct activities in particular sectors, such as education or agriculture.[12] Sector loans in forestry have been a more recent addition, with the idea that forestry projects, to be successful, must exist within a more rational and supportive policy framework than they have in the past (World Bank 1991a, 1991b). Still, most of the Bank's loans are made through projects. If one views the growth of an organization as a measure of its success, the project approach, as the central component of its core technology, has served the Bank well.

Growth and Flexibility of the World Bank's Lending Program

The World Bank is a profit-making organization, although it does not pay dividends to shareholders. To remain credible within the eyes of the conservative financial world, the Bank must maintain its own creditworthiness. The utilization of development projects with sound economic and financial criteria—a good return on investments, a thorough review of the borrower's ability to repay the loan, and, especially, government member guarantees—has given the Bank enormous credibility within the world's financial markets. The Bank has been very successful as a financial institution, with its bonds rated AAA. And as an organization the Bank continues to grow. For example, in 1950 the size of the Bank's annual lending portfolio was $166 million. In 1986 it was over $16 billion, and by 1992 it was nearly $22 billion (unadjusted) (see World Bank annual reports).

The Bank has also through the years expanded its lending activities to include a larger number of sectors. In 1960, for example, 94 percent of its lending portfolio went to projects in three sectors: industry (22%), electric power (dams) (35%), and transportation (roads) (37%). By 1992 these same three sectors composed only 28 percent of the Bank's lending efforts. The largest gain by any one sector during the 1970s and 1980s has been in agriculture and rural development. Agriculture was 5 percent in 1960, hovered around the 30 percent mark from the mid-1970s to the late 1980s, and has since dipped a bit. Other sectors that have emerged over the years include education, population planning and health, urban development, nonelectric-power energy development, and water supply and sewage control. The greatest growth in both the size of the Bank's lending portfolio and the number of lending sectors occurred during the 1970s under the leadership of Robert McNamara.

The Bank's ability to move into new areas has certainly aided its growth and, consequently, its success. The Bank has demonstrated that its technology has some degree of adaptability. It would appear, at least, that the Bank can change to fit the task. In this sense the Bank's technology is a relatively "neutral" one, which can perform a number of functions (see Ness and Brechin 1988), thus allowing the Bank to adjust to changing environmental conditions and making it more likely to survive and prosper.

Fear exists within the Bank that abandoning or modifying the project approach could severely weaken the organization (see, e.g., Ayres 1983; Crane and Finkle 1981; Hawkins 1970). As a consequence, there are internal pressures to maintain the project approach and reject any serious modification of it, although the share of nonproject loans continued to rise during the 1980s. In spite of its adaptive appearance, the project approach has limited the Bank's performance in a number of sectors, including forestry.

Evolution of the Project Approach

The Bank did not invent the concept of the project loan, but it did put it to greater use. Project loans are required by the Bank's charter. The Bank's Articles of Agreement specify that "loans made or guaranteed by the Bank shall, except in special circumstances, be for purposes of specific projects for reconstruction or development" (Article Ill, Section 4, vii). The notion was that the Bank would be a place where governmental officials would come with prepared loan proposals in hand and where the Bank would independently and with great professional skill evaluate each proposal on its own merits, say "Yes" or "No" and "Next, please." This was part of the "arm's length" concept. To maintain its integrity, the Bank was not to become involved in either the proposal preparation itself or its implementation. It was also to be a lender of last resort, when private capital could not be found to finance the project. As the Bank emerged from its shell in the late 1940s, however, it quickly became apparent that it was to be an organization quite different from what its creators intended.

A number of events forced the Bank to take on unintended roles. First, the advent of the U.S. Marshall Plan of the late 1940s, for Europe's reconstruction, closed the door on the Bank's role of rebuilding the war-torn continent, after only four loans were approved. This forced the Bank to shift all its attention to what was thought would be its lesser role—Third World development. It made its first development project loan to Chile in 1948.

Second, and most important to the discussion here, is that few countries in the developing world were able to produce the type of project loan proposals suitable for the Bank to simply evaluate independently. In order to make loans and move its money, it was (and has remained) necessary for the Bank to assist many of its clients in the identification as well as preparation of appropriate development loan proposals. In short, the Bank has become a partner in the loan development process, not just an unbiased lender. As noted by Mason and Asher, (1973) the "arm" in "arms length" quickly became much shorter (256).

Description of the Project Approach

The project approach can be discussed in the context of the project cycle and the discussions and negotiations between the Bank's project staff and the officials from the borrowing country. We should begin with a brief look at the concept of a development project.

Although the Bank's charter declares that only loans for specific projects will be issued, the Bank's framers never defined what they meant by a project. As noted by Baum and Tolbert (1985), "the concept of a project, however, has evolved and the range of activities embraced within it has expanded and become more diverse"(7). Nevertheless, they go on to note that the Bank's development projects tend to embrace several or all of the following elements:

- Capital investment in civil works, equipment, or both (the so-called bricks and mortar of the project)
- Provision of services for design and engineering, supervision of construction and improvement of operations and maintenance

- Strengthening of local institutions concerned with implementing and operating the project, including the training of local managers and staff
- Improvements in policies—such as those on pricing, subsidies, and cost recovery—that affect project performance and the relationship of the project both to the sector in which it falls and to broader national development objectives
- A plan for implementing the above activities to achieve the project's objectives within a given time (Baum and Tolbert 1985, 8).

Other authors (see Ayres 1983, 41; Chadenet and King 1972) provide similar descriptions of the project approach.

With the wide latitude given to it, the concept of a project loan becomes far from a simple or straightforward endeavor. Yet the project concept attempts to retain the loan proposal as a limited and identifiable package that can submit to the rigors of economic and technical appraisal. Perhaps the most identifiable aspect of the Bank's effort may not be the particular project itself but the sequenced pattern that all projects follow.

This pattern, called the "project cycle," comprises several stages. Baum and Tolbert (1985) listed the stages as project identification, preparation, implementation, and evaluation. Ayres (1983) labeled the stages slightly differently: identification, preparation, appraisal, negotiation, and supervision (44). Although both lists are correct, Ayres's is more explicit about the Bank's role in the process.

The Bank's involvement in just about all the stages varies, depending on the specific situation. However, one stage about which the Bank has remained most adamant is implementation. The Bank has never implemented a project and likely never will. This has always been left to specified agencies within the borrowing country or, in certain circumstances, to hired contractors. The Bank's lack of interest in actually implementing projects is a vestige of the "arm's length" concept. Regardless of the amount of technical assistance the Bank provides its borrowers today, it continues to view itself as a development project financier and not an executing agency. The Bank has felt that it would jeopardize its status as a lender too greatly if it implemented the projects it financed (Mason and Asher 1973, 251).

This is not to say, however, that the Bank has been completely uninterested in how the loan is spent (see World Bank 1992b "Wapenhans Report"). But it was forced to compromise on the issue of supervision some time ago. To maintain its image as an "unbiased" lender, yet oversee its loans, the Bank has settled on a supervisory system based on periodic field missions by its staff and systematic reporting by the implementing agency. The Bank has maintained centralized control over the disbursement of project loan funds, which are only released if the project meets the expectations of the staff and if the necessary reports have been filed. Remember, however, that the Bank's prime objective has been to move money. From a sociological perspective, one could go so far as to suggest that the Bank's economic and financial project analyses are equally a ceremonial (sensu Meyer and Rowan 1979) undertaken to maintain its image as a competent lending institution. Hard though it may be to believe, the actual success of a project in meeting development needs has not been an integral concern of the Bank as long as the loan process remains unaffected. Only recently has the Bank itself recognized this focus (see again World Bank 1992b, "Wapenhans Report").

When the project is completed, the Bank staff is required to file a completion report. This is usually done with the assistance of the executing agency. Only some of the projects are actually independently evaluated. These evaluations are conducted by the OED (Operations Evaluation Department), an independent division within the Bank established by McNamara in 1975. To assure its objectivity a bit more, the OED staff reports directly to the Bank's president and executive directors, not to the project staff. OED's staff and budget are small, however; and, augmented by hired consultants, OED evaluates only about half of the 250 to 270 completed projects that the Bank completes each year (Baum and Tolbert 1985, 384; see also World Bank annual reports). These evaluations—or performance audits, as they are called—generally take place one to five years after the project has been completed. Instead of attempting to measure the project's impact on its target population, most performance audits attempt to evaluate the "correctness" of the appraisal report (i.e., Was the blueprint properly designed and then followed?). A critical part of the investigation is to recalculate the benefit/cost ratios. Observers have consistently noted the benefit/cost ratios are, on average and across all sectors, 50 percent lower than expected (Ayres 1983, 45–46; van de Laar 1980, 4; World Bank 1992b, "Wapenhans Report"). This finding supports the argument that even the economic "soundness" of the implemented project is less important than the agreement with the ministry to move the money and to repay the loan.[13]

The heart of the project approach is the project appraisal. The appraisal is both a document and a process. As a document, the appraisal describes the nature of the works to be carried out, the intended beneficiaries, the economic and financial justifications, and the conditions thought necessary for success (Ayres 1983, 45; Baum and Tolbert 1985; World Bank 1991c). The project appraisal document is not an evaluation of the project, as the term might suggest. Rather, it is actually the project proposal, economic and financial justification, and "blueprint" for implementation.

The appraisal document is always written by Bank staff and usually takes more than a year to complete. The Bank takes the ritual of the appraisal process with utmost seriousness. As noted by Mason and Asher (1973), the process is clearly a function of the Bank as a lender (235). Conversations with Bank staff have revealed that appraising projects is what the Bank does best. Although projects need to be technically and socially sound, the heart of the appraisal process historically has been the project's economic and financial analyses. Here the project's economic benefits and costs are calculated along with the client's ability to pay for the loan.[14]

As a process, the project appraisal is an elaborate procedure of internal reviews and external negotiations. The internal reviews are designed to serve as a quality-control check for both borrower and lender before the project appraisal document is formally approved by the Bank. The external negotiations define the nature of the project to be funded and its terms of agreement.

World Bank's Negotiations with Country Officials

The other central component of the project approach is the constant discussions and negotiations held between the World Bank staff and officials from developing countries. The Bank has long been proud of the "continuous dialogue" it maintains with its clients (see Crane and Finkle 1981). Recall that the Bank is required to deal only with the client's treasury or ministry of finance, but usually there are discussions with heads of the implementing agency as well. The loan negotiations are an iterative process until the project is formally approved.

Because of the political sensitivity of refusing loans to members, and the considerable time invested in each project proposal, all disagreements are worked out before a project loan is formally submitted to the Bank's executive directors for approval (Mason and Asher 1973, 186). From their reviews, Ayres (1983) and Rich (1994) concluded that the executive directors have never formally turned down a loan application.

Critiquing the World Bank's Core Technology

Even though it has been the heart of the World Bank's operations, Bank officials as well as external observers have noted that the project approach has not been without limitations. Baum and Tolbert (1985) provide a standard critique of the project approach: "It depends on quantitative inputs of data and can be no more reliable than those data. It also depends on estimates and forecasts, which are subject to human error. Value judgments must be made, but the project approach should at least force them to be made explicitly. Risks can be assessed but not avoided, and projects must be designed and implemented against a constantly shifting background of political, social, and economic changes. In the last analysis, the effectiveness of the project approach depends on the skill and judgement of those who use it" (335).

Although Baum and Tolbert's assessment is a rather benign criticism, the unavailability of certain data has without doubt affected how the Bank has evaluated project loans. Unequivocally, the more hard data, conservative techniques, and forecasts that are used, the more likely that the proposal will avoid criticisms or raise concerns. Whenever data, forecasts, and procedures become questionable, the appraisal process has tended to be biased against that proposed project, regardless of its other merits. This is the crux of the dichotomy between the Bank's roles as a bank and a development agency. The issue over hard data has biased the process away from projects that are more difficult to quantify, such as those focused on poverty, the environment, and other "social" investments (Ayres 1983; McGaughey 1986; Rich 1986, 1994; World Bank 1991c). Forestry projects, and especially rural development forestry, tend to have long pay-back periods and undervalued benefits and often fail to generate desperately needed cash as do many export-oriented projects. These concerns over microeconomic issues of project appraisal reflect the interaction between the Bank's core technology and the nature of the task to be performed.

There are other deleterious consequences of the project approach. Although the approach can be applied to a number of sectors, its performance among those sectors varies considerably. I consider the interaction of the task with the Bank's investment technology to be a key determinant of this variance. The project approach seems to be an extremely appropriate tool for certain types of projects, but not for others. Projects aimed at direct human interactions are served best when flexibility is built into the projects so that they can meet unexpected situations and problems (see Ayres 1983; Clements 1993; Leach and Mearns 1988). Other projects, such as those to build physical infrastructure, have relatively few dynamic problems and require less flexibility in their design. Through the Bank's project appraisal documents, staff have tended to outline the project's course of events—what is to be done, when, and how. In this sense, the Bank's formal procedures have made its project activities linear and inflexible. The project approach has often forcibly fitted the task to the organization rather than the other way around (Ayres 1983; Crane and Finkle 1981; Korten 1980). Examples can be provided from programs in poverty alleviation, population control, rural development, and forestry. Take, for example, the interview I had with the former director of Senegal's Forestry Department. He described his experiences with the Bank. He was most impressed with the Bank staff's ability to find creative financing solutions to assure that the proposal would be accepted. He was, however, most disappointed in how rigid the staff was in its interpretation of how the project was to be carried out. "Everything had to be followed just so," he recalled. This rigidity did not allow the project to be changed to redress unforeseen problems that emerged during its implementation.

Because the Bank created the project appraisal document, in part, as a means of controlling project quality and accountability, the Bank's approach to development has necessarily been top-down. The Bank staff, in consultation with high-level government officials from the client country, directs the course of project events from afar with a varying but relatively high degree of abstraction. Although the Bank staff may have made a number of appraisal trips to gather important information about the proposed project, most, if not all, the projects that deal with the human element are much too complex and uncertain to be designed in a fail-safe manner. Although the Bank's top-down, blueprint approach fits its boarder organizational goals, this approach does a disservice to those projects that should be designed more from the bottom-up and that require greater flexibility in their implementation. This problem is compounded by the Bank's centralized structure. The Bank does not place project staff in the field to work out problems that arise after the project has begun.

Even though the Bank has expanded into "soft sectors," education, population planning, poverty alleviation, it has, until very recently in some poverty programs, failed to adjust its project approach largely because of the institutionalization of the centralized, formalistic approach. Performance in these soft sectors has suffered accordingly (Ayres 1983; Clements 1993; Crane and Finkle 1981; Korton 1980; van de Laar 1980). And for reasons not entirely clear, at the time of my writing this book the Bank had not moved away from this more rigid approach for its work in forestry, including social forestry.

Another problem with the Bank's project approach has been its bias in favor of large project loans requiring minimal staff time (Ayres 1983;

van de Laar 1980). This attitude stems from the Bank's objective of moving money. (Tendler [1975] also notes this tendency for all large funding agencies.) It is much easier to move money through large infrastructure projects that require relatively little staff time than through the typical "poverty-oriented" projects that tend to be smaller in size, and that require greater commitment of staff time and greater skill to generate and supervise (Ayres 1983, 10). During McNarama's administration there was considerable tension over the need to move more money, as the lending program increased dramatically, and the need to do more for poverty and human welfare. This proved to be an organizational contradiction. This problem can be generalized to any relatively small project or to those that are more staff intensive. This point has held for lending in forestry as well, and especially for community forestry. Because of their characteristics and limited institutional resources, forestry projects—and even more so, rural development forestry projects—have tended to be relatively small in size and to require more staff participation.

A final critique of the project approach concerns the Bank's role as financier versus implementor. Since the Bank does not implement the projects it finances, the actual results are in the hands of others—generally agencies within the client government. Although the Bank attempts to supervise, it does so mostly from a distance through occasional field trips and written reports. Consequently, it is easy to see why a number of Bank projects have problems in their implementation. Projects pursued by governments are approved more on the basis of their financial needs and less on the technical abilities of their executing agencies, which are often lacking in resources of all kinds (see also the discussion of the Morris Commission in Rich 1994). Some governments do hire independent contractors to carry out the projects. Although often competent, the contractors too have been constrained by the Bank's rigid criteria. It seems ironic that the Bank has refused to step over the line from financier to implementor even though it sees the need to help client governments prepare proposals and is fully aware of the problems its approach has generated. This point clearly reflects once again the Bank's central characteristic as a mover of money, which it can pursue while distancing itself from what actually happens on the ground in the name of national sovereignty.[15] Becoming an implementor would make the Bank even more responsible for project outcomes and could jeopardize its lending operations. Ironically, its lack of interest, too, over the environmental and social impacts of its projects has now begun to jeopardize its lending operations—this time because of severe criticism from numerous environmental and human rights groups, some of which are calling for the Bank's dismantling.

To summarize this section, several characteristics of the Bank's core technology stand out as having important effects (1) on the *type* of projects supported internally by the Bank and supported externally by the client and (2) on the project's overall *performance* or impact. Internally, the Bank's technology has been predisposed to select those types of projects that are large and for which the staff can more easily calculate positive benefit-to-cost ratios. Export-earning and infrastructure projects have been favored over human development projects. The Bank's top-down, blueprint appraisal process has generally made its projects inflexible—in turn, making bottom-up projects more difficult.

Externally, the Bank's clients obviously like to select projects that enhance their own agendas. They have tended to select those projects that have limited time horizons and that are export generating. Tree-planting activities for rural communities have historically fallen short on both accounts. In addition, if the executing agency is small or poorly organized, it will not have sufficient capacity to absorb large loans, thus making the project less attractive to the Bank and government; the agency will also not perform well, thus hindering development. But a failure to properly implement a project may not be wholly or even primarily the fault of the executing agency. The Bank's blueprint approach may constrain the agency too much. Still, the client country's ministry of finance may have approved a project simply to obtain the desired financial resources without much consideration of the needs and limitations of the executing agency in properly implementing the loan. Baum and Tolbert (1985) list the failure of executing agencies to adequately implement projects as the principal reason for the failure of Bank-financed projects. Improper project design was listed as the second most common reason. For forestry projects, the Bank's own evaluation in 1991 found numerous problems in administering agencies, but especially in the Bank's inability to place the project within the proper policy framework or to incorporate complex social conditions into its designs (World Bank 1991b).

Chapter 2

Of Diplomats and Foresters: The Food and Agriculture Organization

Since 1945, the Food and Agriculture Organization (FAO) has become the largest of the United Nations' specialized agencies, and, as its name denotes, it concentrates on matters of food and agriculture, including fisheries and forestry. Unlike the World Bank, FAO does not make loans to member governments; rather, it provides technical support to those who request it. Like the Bank, FAO cultivates the international "intellect" on the matters of its domain. Its environmental forces, however, significantly shape its activities and are central in defining the key differences between it, the Bank, and CARE.

FAO performs four basic functions, as: (1) a development agency that carries out technical advice and assistance on behalf of governments and funding agencies; (2) an information center that collects and disseminates technical knowledge; (3) an adviser to governments, especially regarding policy and planning issues; and (4) a neutral forum where governments can gather to meet and discuss relevant matters. FAO's Forestry Department attempts to fulfill these basic functions as the international community's agency for forestry matters.

FAO Forestry has nearly separate dual structures. Generally speaking, the more centralized Regular Programme deals with more abstract concerns of policy and informational needs. The Field Programme is largely a decentralized structure that assists individual nations with their more-on-the-ground forestry needs. FAO Forestry is essentially a service agency, created to cater to member nations. As a service agency for the United Nations, it is highly dependent upon others for financial support and policy direction.

Precisely because of these dependencies, its core technology of professional forestry knowledge and skills is much more adaptable and flexible than I originally thought. It is essentially a diplomatic agency that, paradoxically, can only lead when following. Although FAO Forestry's movement into community forestry was shaped by changes in the larger institutional environment, FAO Forestry was also heavily influenced by the desires of particular players within its task environment and had to to adapt its technology to reflect those desires.

The Development of Community and Rural Development Forestry

FAO Forestry's work in community and rural development forestry began formally in 1979 when its Forestry for Local Community Development (FLCD) program was established. FLCD came to life with an exchange of letters between the Government of Sweden, funder of the program, and FAO, dated October 24, and November 6, 1979, respectively (FAO 1987, 3). The Government of Sweden continued to be the only support for the FLCD program until 1986. In 1987 the program's name changed to Forests, Trees, and People Programme (FTPP) with both a broader mission and trust fund support.

Although the program was formally established in 1979, the Swedish government, through the Swedish International Development Authority (SIDA), had been working with FAO on community forestry issues since the early to mid-1970s (Brechin 1989, 191). They had cosponsored a number of community forestry activities, such as three expert consultations, two FAO publications, three mission presentations, and three background publications as a means of testing the political and technical waters.

Processes of Institutionalization

It is clear that SIDA played an important role in helping FAO Forestry develop expertise in community forestry. SIDA's involvement, however, was more a proximate than ultimate cause. Like the World Bank, FAO was primed to endorse community forestry by a series of external events that shaped its larger cultural environment. Four events stand out: (1) the United Nations Conference on the Human Environment, held in Stockholm in 1972; (2) the World Conference on Agrarian Reform and Rural Development in 1980; (3) the Eighth World Forestry Congress, which took place in Jakarta in 1978; and (4) the World Bank's interest in promoting rural development and assistance to subsistence farmers, which was first articulated in the mid-1970s.

The United Nations Conference on the Human Environment

Stockholm was the site of the first large international conference on the biophysical environment, essentially internationalizing the movement that had already gained prominence in the West. The conference not only brought concern for the environment to the international stage but, more important, attempted to link that concern to the general social and economic welfare of the Third World (Caldwell 1984, 1990). The conference was instrumental in redefining forestry's role beyond timber production to environmental protection, although FAO documents associated with the conference made no direct references to community forestry.

FAO Forestry was asked to contribute its expertise to certain aspects of the conference agenda. In particular, it collaborated with UNESCO and the World Meteorological Organization (WMO) in producing studies on forests and with UNESCO and IUCN (International Union for Conservation of Nature and Natural Resources) in producing studies on wildlife and national parks.[1] FAO Forestry's studies stressed the need to better monitor and manage the resource base and to develop better mechanisms for exchanging information on environmental degradation.

The conference did, however, widen the scope of forestry matters. Instead of an isolated resource concern, forests were viewed more as an integral part of large land-use systems:

> Forests are subject to increasing pressure by competing forms of land use and from commercial utilization. Some areas are better able to cope with these conflicting demands than others; in tropical regions, arid regions, regions of dense population or heavily industrialized, forest depletion and degradation are taking place at an accelerated rate. Forest management should provide for the maintenance of both the productive capacity and protective role of forests, and give recognition to their value in the protection of other natural resources and in the enhancement of the environment. Forest legislation and institutions should be adjusted accordingly.[2]

What is absent from this quotation, and from the material from which it was drawn, is any direct reference to rural populations and local forestry needs, such as fuelwood or agroforestry. This remained true in 1974 at the second session of the Committee on Forestry (COFO), FOA Forestry's Standing Committee to the FAO Council, or governing body. There, COFO approved of FAO Forestry's activities of the past two years. It espoused essentially the traditional values associated with traditional forestry management.

> The Committee endorsed the philosophy of FAO that forestry must mean the management of forest lands to provide the optimum return in human benefits, on a sustained yield basis, while at the same time ensuring the constant improvement of the resources. These benefits to society include not only tangible commodities such as timber, forage and minor forest products, but also services in the form of protection of soil, flora and fauna, and the satisfaction of the increasingly important recreational and psychological needs of man. Thus the Committee underlined the importance of forestry's role in the ecological management of natural resources.[3]

In the paragraph that followed this quotation, however, there was a sentence suggesting that because forestry now is seen as having a broader environmental role, *"the participation of the public is needed"* (emphasis added) to help determine the goals of forest management.

The Stockholm conference did not raise directly the concept of community forestry. Rather, it planted the seed by expanding the view of forestry to include a larger environmental focus. It was not until COFO's third session, in 1976, that direct reference to community forestry was made. At that meeting, the FAO Forestry Secretariat stated:

> Present forest policy and use of forests as they are currently applied in most developing countries aim primarily at the mobilization of the capital in mature or over-mature natural forests and in some cases at the building up of large industrial plantations of fast growing species as raw material for wood-based industries. [This] paper shows that this policy may well contribute to the overall development of the national economy but is not adequate to fight against the trend of further impoverishment of remote rural areas. A new dimension of forestry is therefore needed which can contribute to stabilizing the natural foundations of food production and to stop or even to reverse the impoverishment of rural areas. The so-called "Forestry for Local Community Development" implies a special type of forestry allied for, by or on behalf of a local community such as a village, a group of villages or a number of individual settlements. The objectives of the management of such forests are primarily the production of goods and services to cover the needs of the local community and their population.[4]

WCARRD and the Eighth World Forestry Congress

In addition to the WCARRD (World Conference on Agrarian Reform and Rural Development) discussed later, another source in the institutionalization of community forestry at FAO and the world community at large was the Eighth World Forestry Congress, held in Jakarta in 1978. This conference specifically focused on the human element as illustrated by its official theme, "Forestry for People." It was the first large international forum where the concepts of community and rural development forestry were discussed. The Congress's subject matter was so well received by member countries that it influenced the proceedings of the ensuing twentieth session of the FAO Conference in 1979. Fontaine (1985) declared that Jakarta marked a turning point in the history of forestry (10). Forestry became a more appreciated and essential ingredient in the economic and social welfare of developing nations. Community forestry and forestry for rural development were formally established in the political arena. A longtime FAO official and early advocate of people forestry, Jack Westoby, resigned from FAO Forestry in 1975 to promote the 1978 World Forestry Congress's theme from the outside where he thought he could be more effective (Westoby 1988, personal communication).

The environmental and social realities of many developing countries finally forced the entire world to look differently at the role of forestry in the development process. Forestry from that perspective was thought to encompass new and broader economic and social functions. Forestry now was to touch the daily lives of billions of poor rural people. The kind of forestry that had a narrow, industrial focus, characterized by timber production and pulp and paper industries, became only one type of forestry. Forestry for the people and communities became the other type. Although the Jakarta World Congress certainly legitimized the notion of people forestry, this event itself was an outcome of the same paradigm shift that influenced the Bank's endorsement of community and rural development forestry.

The Influence of the World Bank

In the documents from the second session of COFO in 1974, there was a small paragraph noting a recent meeting between FAO Forestry and World Bank officials. The meeting focused on the Bank's newfound interest in funding projects geared to improving the conditions of subsistence farmers. It was noted that UNDP was contemplating a similar program. FAO Forestry responded by stating: "It is therefore highly likely that it will become necessary for the Department to evidence the contribution of forestry to the welfare of the subsistence farmer, especially in the least developed nations."[5]

The brief report on the Bank's discussion with FAO Forestry was the first time (in FAO documents at least) that small farmers were linked to the concept of rural development. In the document, FAO Forestry noted that it would investigate the best approaches for the department to pursue activities in this problem area.

The meeting preceded by a few months the Bank's *rural development sector policy paper,* which was published in February 1975. As discussed in Chapter 1, the sector policy paper formally announced the Bank's intention to fund rural development projects, particularly aimed at small farmers. It is likely that the Bank's interest in rural poverty helped to spark interest at FAO and to provide greater legitimacy and rationale for the former's movement into community forestry.

From a broader perspective, the Bank's movement into community and rural development forestry was part of a larger general trend in the theory and practice of economic development (as was more fully discussed in chapter 1.) *In this sense, FAO's movement in this new direction was also only a product of that larger trend.* It also indicates that FAO was a tad slower than other organizations in formally recognizing that trend.

SIDA and FAO Forestry: Development of Professional Knowledge in Community Forestry

Given the events presented above, SIDA (Swedish International Development Agency) played an important but only a proximate role in moving FAO Forestry toward community forestry. Drawing upon resources found within the larger institutional environment, SIDA was able to push FAO Forestry into this new type of forestry by providing financial backing. SIDA has had a long history of funding forestry projects through FAO Forestry. One former FAO Forestry official noted that the "content of this joint effort was continuously evolving" (Arnold 1988, personal communication). At that time, both SIDA and FAO Forestry, as well as many other organizations, were moved by the growing problems of desertification and fuelwood scarcity, especially in the Sahel (Arnold 1988, personal communication). Both saw the need to reorient forestry in order to meet the needs of rural communities (FAO 1987, 3). As a consequence, three expert consultations were held between October 1976 and December 1977 (FAO 1987, 72). As a product of this effort, an important publication was released in 1978 entitled

Forestry for Local Community Development (FAO 1987, 72). In addition, in late 1978 and in 1979, three workshops on community forestry were held in the Sudan, Senegal, and Paraguay (FAO 1987, 72). With SIDA's financial support and constant encouragement, FAO Forestry was able to develop professional expertise in community forestry. With this groundwork established, SIDA funded FLCD (Forestry for Local Community Development Programme), in July of 1979. FAO Forestry became an agency directly concerned with the welfare of rural peoples.

Although SIDA's involvement has been considered important in FAO Forestry's movement into community, one former FAO Forestry official closely associated with the FLCD program noted that SIDA was not irreplaceable in that effort. He argued that if SIDA had not stepped forward to promote community forestry within FAO Forestry, some other organization would have done so, since the concept was rapidly diffusing through the general environment (Arnold 1988, personal communication). Nonetheless, SIDA's involvement was unique. Instead of working through FAO's Field Programme, which is the typical approach of donor members, SIDA chose to develop FAO Forestry's Regular Programme capabilities in community forestry directly, thereby institutionalizing community forestry within the FAO Forestry Department itself (Westoby 1988, personal communication; Arnold 1988, personal communication). This approach was probably responsible for the profound changes in FAO Forestry's technology and character that resulted in the organization's embracing community forestry. *The important point here, however, is, that it took an outside organization (donor member) to move FAO Forestry in a new direction.* This circumstance demonstrates the constrained nature of FAO's service agency character and provides a glimpse into how larger cultural changes can become realities within specific organizations.

As in the previous discussion of the Bank's organizational character, to present a fairly full view of FAO Forestry, it is once again necessary to briefly review its birth, mission, and history.

The Character of FAO's Forestry Department: A Brief History

Forestry was an afterthought in the establishment of FAO. In the spring of 1943, the initial meeting that led to the eventual creation of FAO was held in the United States at Hot Springs, Virginia. At the Hot Springs Conference, there was considerable resistance to forestry's becoming a part of FAO's mandate, thus reflecting the historic tensions between forestry and agriculture. The agriculturists won the day, and forestry was intentionally excluded from FAO's terms of reference established at the conference (Myrdal 1978; Phillips 1981; Winters 1974).

Following the conference a number of determined foresters gathered in Washington, D.C., to pursue the matter. They were able to persuade a number of influential individuals of the importance and relevance of forestry to the mission of food and agriculture. Among those whom they convinced were Lester B. Pearson of Canada, Chairman of the Interim Commission of FAO; Dean Acheson of the United States, then Assistant Secretary of State; and Frank L. McDougall of Australia, who was a respected member of the Interim Commission. The matter was eventually taken up with President Franklin D. Roosevelt, who on November 2, 1943, personally approved the inclusion of forestry in FAO's terms of reference. Roosevelt's wish was immediately accepted, given the unprecedented level of U.S. prestige after the war (Myrdal 1978; Phillips 1981; Winters 1974).

The Interim Commission under Pearson quickly estalished an advisory committee called the Technical Committee on Forestry and Primary Forest Products. The committee documented the close interrelationship between forests and agricultural production, as well as forestry's ability to add directly to the expansion of the world economy (Phillips 1981, 141). The committee's final report was one of five that were collectively published as the Interim Commission's own final report: *Five Technical Reports on Food and Agriculture* (1945).[6] The commission's report, presented at a conference in Quebec in 1945, provided the basis for FAO's future programs. In October 1945, FAO was officially established, with forestry in place as a part of its program of work. From that uncertain beginning, forestry within FAO has slowly grown to become an important and integral part of its program of work. This growth culminated in forestry's achieving FAO departmental status in 1970.

The Development of the Forestry Program

FAO Forestry's original mission was "to supply an expanding world economy with useful forest products in the qualities and quantities adequate to sustain the general welfare" (Leloup 1985, 15). To achieve this general goal, four objectives for forestry were devised:

1. Increase the yield from forests now being used.
2. Reduce waste from logging operations and in the conversion of timber into products.
3. Open up virgin forests.
4. Plant new forests.

It was not until April 1947, however, that FAO Forestry had obtained a formal structure to pursue those aims. The Forestry and Forest Products Division consisted of only two branches: the Forestry Branch and the Forest Products Branch. From this time through January 1970, when it achieved departmental status, forestry within FAO utilized a number of different titles.[7] But regardless of its name, for its first 35 years industrial forestry dominated the work and thinking of FAO Forestry.

Through the years, FAO Forestry experienced a number of changes in its program of work. The most notable change in the early years was the development of its technical assistance program (i.e. field projects) in the 1950s. This program blossomed with decolonization and the proliferation of new member countries in the 1950s and 1960s. And in turn, FAO Forestry's movement into Third World technical assistance altered the organization's character and operations.

The most distinctive change of the 1960s and 1970s, however, was probably in FAO's view of the development process. Concern today for the development of rural areas in the Third World was also overshadowed by the emphasis on the urban, industrial, and export market development.[8] The new focus on rural development changed perceptions of forestry and forestry problems. *The planting and management of trees became a concern of the rural poor, as well as of national governments and timber corporations.*

Reflecting the changes that have occurred, FAO Forestry's stated aims have also changed. In 1981, an appropriate time frame for this discussion, FAO Forestry's aims for assisting its member countries were as follows:

1. Increase the outputs of goods and services from their forests while maintaining production capacities.
2. Expand socioeconomic benefits from the use of forest resources, including the contribution of forestry to the food security.
3. *Achieving an equitable distribution of such socioeconomic benefits in rural areas.* (emphasis added).[9]

When compared with the stated aims of 1945, those of 1981 show a much more noticeable interest in the welfare of people in rural areas and in the expansion of forestry issues to the broader concerns of socioeconomic development.

In his personal reflections on FAO Forestry's fortieth anniversary in 1985, the former director of its Forests Resources Division, R. G. Fontaine, organized FAO Forestry's history into three distinct periods: 1945 to 1959, 1960 to 1969, and 1970 to 1985 (Fontaine 1985).

FAO Forestry, 1945–1959

During its earliest years, the staff at FAO Forestry concerned themselves mostly with building the organization itself, developing its program; the rebuilding of war-ravaged Europe; and collecting global data on forest supply and the demand for forest products. The need for greater information on existing forest resources and for the protection and management of the vast undeveloped forests of tropical regions dominated the forestry-related discussions at the first FAO Conference in 1945 (FAO 1985a). The information collected from the conference led to the publication in 1948 of *Forest Resources of the World.* There were three subsequent editions over the next 15 years. The information was essential in establishing forest policies throughout the world, which largely called for the creation or development of forest agencies within governments to manage this vast resource. It was also the theme of the Third World Forestry Congress in Helsinki in 1949 (FAO 1985a, 55; Fontaine 1985, 7).

FAO Forestry, during this era, worked intensively on regional issues, with an overwhelming emphasis on Europe. In 1947, FAO Forestry organized a conference at Marianske-Lanzne, Czechoslovakia. The conference assisted the United Nations Economic Commission for Europe (ECE) in working out crucial timber trade agreements between Eastern and Western European countries. The agreement allowed for a greater flow of much-needed timber into Western Europe to aid in postwar reconstruction (FAO 1985a; Fontaine 1985). FAO Forestry and ECE collaborated once again in the 1950s on the European trend studies. These studies, which served as models for other world regions, looked at timber consumption and production and trade trends for Europe. In addition to Europe, forestry commissions for promoting industrial forestry were formed in Latin America (1949), Asia (1950), the Near East (1953), Africa (1959), and North America (1959).

Like the other international organizations at the time, FAO Forestry focused almost entirely on the collection and dissemination of information of a truly international nature. Most of its activities consisted of international conferences and other meetings, as well as the publication of technical reports and studies on regional and global forestry matters. The goal of these activities was largely to promote forestry as a profession and its role in general economic expansion of all nations (see FAO 1985a; Fontaine 1985; Phillips 1981).

These types of activities of FAO Forestry in the period 1945–50 formed the traditional and ongoing program of the organization. The activities would eventually compose the Regular Programme, which has been financed from member-country dues.

In the late 1940s and early 1950s, a new program concept emerged in FAO, called Technical Assistance, or, sometimes, Technical Cooperation. In addition to the international staff's working on more regional or global matters, they also began to provide expert advice to individual member countries that requested assistance. Thus was born the UN's field assistance programs. This innovation rapidly diffused throughout the UN system. The first technical assistance officer for FAO Forestry was appointed in 1951 (Phillips 1981, 47).

The funding for the field program came from special UN allocations. The first source was EPTA (Expanded Program of Technical Assistance), which started in 1951 and provided monies for the technical experts who were placed in the various UN agencies. In 1958, another field program funding source emerged. The United Nations Special Fund (UNSF) was created to provide monies for larger and more elaborate fieldwork, such as development projects and research similar to the field programs that exist today (Phillips 1981, 72; Westoby 1975, 208).[10] With these two funding sources, technical assistance flourished in the 1960s.

FAO Forestry: 1960–1969

For the most part, 1960–69 saw a continuation of the regional and international activities developed in the first period (Fontaine 1985, 9). There were, however, several events that distinguished the period. One was the shift of the dominance of forestry by the North Atlantic countries to the countries of the developing world.[11] Another was the rapid growth in UN membership that followed decolonization. As a consequence, FAO and FAO Forestry membership grew rapidly. In 1945, at FAO's first conference, 39 countries joined the organization. By 1969, it had 117 members, and by 1992 membership had risen to 169 (Saouma 1993). The arrival of many new but poor nations led to a greater demand for technical assistance and focusing on Third World forestry. Along with the growth in new nations, came greater demand from these countries for assistance to help them develop modern economies. (This key development reflected changes in the larger institutional environment that affected all three of the organizations discussed in this book. It is a topic to which I return later.) Another event of the period was the development of new administrative structures and a number of cooperative efforts among the international organizations to help provide that assistance.

During this period there was also a shift in the nature of the field activities. They were broadened to include operational and field projects beyond the simple presence of technical experts who provided advice (Fontaine 1985, 9; Muthoo 1985; Westoby 1975, 208). This was an important change and reflects the type of field projects that are presently under way. It was a move away from advice that generally flowed one way to more collaborative field activities. It was a change from prescription to participation and from advice to action.

To meet the increasing need for technical assistance, a number of administrative developments occurred. EPTA (Expanded Program of Technical Assistance) and UNSF (United Nations Special Fund) consolidated in 1965 to become the United Nations Development Programme (UNDP) (Muthoo 1985; Phillips 1981: 73). UNDP became and has remained the largest single funder of FAO's field projects. The increased demand for technical assistance also began to burden the Regular Programme staff. Within FAO, a Development Department was created in 1968 to handle the organization's Field Programme (Phillips 1981, 73). Similar changes took place within FAO Forestry with the expansion of the Operations Office to the Operations Service in June of 1968 (Phillips 1981, 147).

Other administrative developments of interest involved the formal collaboration of FAO with other UN agencies. One of the most important affiliations to our review of the forestry program was the FAO and World Bank Cooperative Programme. The Cooperative Programme, which was created in 1964 and fully implemented in 1968, attempted to combine the technical knowledge of FAO with the financial resources of the Bank (FAO 1985b, 12–13, Phillips 1981, 72, 149). This innovation allowed these organizations to develop the investment potential in forestry and forest products for a number of member countries. Many of these efforts led to the preparation of more technically sound forestry-related development projects funded by the Bank and other multilateral funding agencies.

FAO Forestry: 1970–1985

During the year, 1970–85, FAO Forestry again, underwent considerable change this time moving toward a broader concept of forestry. The period began with forestry's receiving departmental status within FAO. Forestry's growth in stature reflected the expanding sense of forestry's contribution to the process of economic development in the developing countries. The first two periods of FAO Forestry activities just reviewed could be categorized generally as times when the development of the forestry sector was the principle objective. During this last period, however, forestry became more closely associated with the process of development. FAO Forestry's objective, then, changed from *the development of forestry to forestry for development* (FAO 1985a, 56).

Several events highlight this important change. One was FAO's sponsorship of the eighth World Forestry Congress in Jakarta, Indonesia, in 1978. Its theme, "Forestry for People," publicly presented a new and revolutionary view of forestry. Forestry was to be geared to the development needs and desires of rural people throughout the developing world. As I indicated in the Introduction, forestry had previously been a more technically narrow profession geared almost entirely to the production of trees for industrial uses. The Jakarta Congress's theme was influenced, in part, by the environmental realities of the late 1960s and early 1970s, which were highlighted at the 1972 Stockholm Conference. FAO Forestry's movement into rural development forestry illustrates well how larger institutional change emerging from the Stockholm Conference became translated into specific programmatic exercises. These meetings revealed a growing awareness of the biophysical environment and how its degradation affects the physical and economic well-being of rural (as well as urban) populations of the Third World.

Another key event was the World Conference on Agrarian Reform and Rural Development (WCARRD). This conference, sponsored by FAO in 1979, established the importance of, and an action program for, encouraging rural development and formally committed FAO to that goal (Arnold 1992; FAO 1983; Saouma 1993). It was an important legitimizing milestone. For forestry, this conference, as well as the Jakarta Conference in 1978, emphasized the broader social responsibilities of the profession in the achievement of rural development (FAO 1983, 37). The conference helped to broaden the concept of economic development to include rural areas and their poor people. With regard to forestry, WCARRD supported the critical role trees play in promoting sound rural development beyond the traditional notion of industrial forestry and soil and water management (FAO 1983, 37). WCARRD helped create a new cultural view of economic development and established an organizational mandate for rural development forestry. In particular, it supported and sanctioned the community forestry concept, which had been partially established within FAO Forestry a year earlier with Swedish backing.

A third event was the creation of FAO Forestry's FLCD (Forestry for Local Community Development) in 1979. This program, sponsored by SIDA, more than any other single event signaled the fundamental change in FAO Forestry's organizational character. It represented as well a radical shift in its core technology. Although some interest and effort in community forestry at FAO dated back to the early 1970s, the creation of FLCD marked the formal institutionalization of the community forestry concept as part of FAO Forestry's program of work.

Reviewing Fontaine's three historical periods clearly shows the changes that have occurred within FAO Forestry. In particular, there was a shift in attention from the needs and concerns of the Western industrialized countries to those of the Third World. This is seen in the growth of FAO Forestry's work in field programs and shifts to concerns of rural development. This shift in focus to Third World rural development needs, however, is the result of FAO Forestry's following broader political and substantive trends found in its larger environment. It was, however, not limited to FAO Forestry. The review of the three periods also reveals FAO Forestry's role as an international diplomatic organization for forestry concerns. FAO Forestry works with governments on policy matters. It collects and disseminates international information, services international conferences, and assists in the negotiation of trade agreements between countries. Thus, if FAO shares the development function with CARE and the World Bank, it is this *diplomatic* function that most distinguishes FAO Forestry from the other two.

The Present Period: Global Assessments and TFAP

FAO Forestry from 1985 to the time of my writing this book has continued in its general service orientation as outlined above, including its work in community and rural development forestry, with greater emphasis on the role of women in forest management issues (see FAO Council 1990). During this present period two events have stood out. One has been FAO Forestry's work in determining the status of tropical forest resources. The other is its central role in TFAP, the Tropical Forest Action Programme.

In preparation for the 1992 United Nations Conference on Environment and Development (UNCED) in Brazil, FAO Forestry, with sponsorship from the European Economic Community, undertook its second global assessment of tropical forests: the Forestry Resources Assessment 1990 Project. With growing international concern about the rate of tropical deforestation, FAO Forestry has worked arduously to dispense information to its members and the world community by providing an assessment of the changing status of tropical forest resources. Surprisingly, the precise extent of tropical forests and their rates of deforestation have been, until recently, poorly known. FAO Forestry has been coordinating efforts over several years now to both understand the state of tropical forests and to institutionalize more systematic assessment procedures. The 1990s show movement toward globalizing the assessment process in order to make the analysis strictly focused, not just on tropical forest resources, but on all forests, including those in temperate climates (see FAO Council 1990, 1993).

FAO Forestry's work in TFAP illustrates well its efforts to closely cooperate with other organizations, such as the World Bank, UNDP, and WRI, to provide services to its members. The point of TFAP was to provide greater rationalization and coordinated support for the forestry sectors in LDCs with the goal of sustainability. From its inception, TFAP was to be a country-led effort. In other words, it was a voluntary program in which interested developing countries more or less controlled the TFAP process for their own interests. FAO Forestry became the coordinating agency for TFAP to assist LDCs with their particular interests and to match these interests with donor countries and agencies.

TFAP has become an enormously controversial enterprise, to say the least, with many criticisms leveled at it. Although the debate still rages on want went wrong and who is responsible, many donor countries and funding agencies, for example, have become dissatisfied with TFAP's and FAO Forestry's poor performance. Some experts claim that the experience has severely damaged FAO Forestry's political future, with a few calling for the creation of a new international forestry agency. At the same time, many developing countries have been generally pleased by the TFAP process because it has helped to develop their forestry sectors, even though there has been some dissatisfaction over the lack of funding from donor countries and agencies. Although there are many aspects to this controversy, a key facet has been the different notions of performance. Donor countries and agencies were hoping to promote the concept of sustainability largely through the protection of biological diversity by preserving tropical forests. The LDCs, on the other hand, were looking at TFAP as a way to provide greater development through a more intense, yet rational and sustainable, use of their forest resources. FAO Forestry, given its nature, could only attempt to appease both sides—a difficult, if not impossible, position. The TFAP process has greatly increased the FAO Field Programme's workload and its sources of revenue. At the same time, this increase has placed unusual burdens on the Regular Programme staff who provide technical backstopping to field projects in a time when budgetary support has declined (FAO Council 1993). Still, the TFAP workload and controversy have seemed to have had little negative effect on its community forestry work, in particular the efforts of its FTPP.

FAO Forestry's Staff

FAO is a large bureaucratic organization. In this sense it is more like the World Bank than CARE USA. In 1981, 7,426 people were employed by all of FAO, with roughly half (3,827) located in Rome (Phillips 1981, 79). By 1992, there were a total of 6,136 employees. Again slightly more than half were located in Rome (FAO 1993). At FAO Forestry, however, the situation is much different. It is a considerably smaller organization, closer to CARE in size and atmosphere. In 1992, there were about 70 relatively "close-knit" professionals located at FAO Forestry headquarters in Rome, with about 250 overseas. The professional forestry staff in Rome was located in a separate building with FAO's Fisheries Department until 1993, several kilometers from FAO's main compound. This created a less hectic and more informal atmosphere then at the main compound, which seemed to suit the FAO Forestry staff.

The quality of the FAO Forestry staff, however, is considerably mixed. Informants I interviewed gave the staff of FAO Forestry's Regular Programme very high marks. But the informants were considerably more cautious in praising the staff of FAO Forestry's Field Programme. Here there was consensus on the tremendous variation in merit. Although politics and issues of representativeness influence the hiring practices of almost all UN agencies, the qualifications of FAO Regular Programme staff remain high.[12] The staff is also well paid, attracting the attention at least of highly qualified personnel. Almost all professional positions at FAO Forestry are held by trained foresters.[13] Most higher-level officials have at least 10 to 15 years of international field experience, which adds practicalness to their academic training.

There is also a "measured" quality about the staff, which reflects its diplomatic character. One FAO Forestry official remarked to me that the staff gets paid for its "brains" and "work" and added, "We are not zealots on a mission." Given the constrained (service agency) character of FAO Forestry, it makes only sense for the agency to hire highly qualified people since a competent staff is the agency's only strength. "Who would seek advice from idiots?" a highly placed FAO official said to me. There is considerable pressure to internationalize the staff. The Regular Programme in particular has been staffed with Westerners, mostly Europeans.

Another issue of the FAO Forestry staff centers around the dual nature of the organization. The staff members of FAO Forestry's Regular Programme and its Field Programme differ in important ways. One way is in size: the Field Programme staff tends to be much larger, two to three times larger than that of the Regular Programme, reflecting FAO Forestry's substantial project work in developing countries. Another way is in staff stability: the size of the Regular Programme staff is fairly stable compared to that of the Field Programme, which fluctuates, often dramatically, with the number and complexity of the field projects funded. A third way in career orientation: The Regular Programme staff consists of career (or career-path) employees (i.e., permanent staff), whereas the Field Programme staff has increasingly been composed of outside experts hired temporarily.[14] And finally, as already indicated, the quality of the overseas staff tends to be more varied. To its advantage FAO Forestry can draw on forestry experts from around the world. It has the potential to select from a wider range of top forestry specialists than does any other forestry-related organization does.[15] It also faces a complicated situation of creating the right match among professional, problem, and country context. When the match is a good one, then there is the potential for a successful project. If the match is not a good one, the opportunity for success is greatly reduced. It is important to note that FAO-hired field professionals usually work for the host government's implementing agency only as technical advisors to the project. All authority lies with the administering agency as negotiated with the funding agency. Consequently, the interaction between FAO and agency staff is essential to the success of the project. In one illustration given to me, FAO sent a forestry expert from South Asia to assist in a project in West Africa. The project quickly unravelled, largely because neither the government officials nor the FAO consultant could get along with one another. Not only does there need to be a solid match between training and experience, given the problem, but between personalities as well. The issue of relationship among FAO, agency staff, and funding institution is discussed more in later chapters.

FAO Forestry's Structure

Unlike the World Bank, which structures its work activities by geographic regions, or CARE, which does so both by regions and technical specialty, FAO divides its work by sector (agriculture, fisheries, forestry, etc.). Because of the nature of its work, FAO Forestry has a dual organizational structure, which I discuss below. To more fully understand the activities and character of FAO Forestry, one must recognize that FAO Forestry is only a small unit within a much larger organization. And, more important, it exists to serve member country governments. This makes it foremost a political organization, but one that provides specific technical services. These characteristics greatly restrain the actions of FAO Forestry. It is influenced in particular by its formal relationships with FAO, as a whole, especially its governing bodies, the FAO Conference, the FAO Council, and COFO.

The FAO Conference

The FAO Conference, which meets every second year, is the supreme governing and deliberative body of FAO, controlling organizational policy and outlining expenditures. It is composed of member-country delegates. Only they may vote on issues, and each has only one vote. This makes FAO a much more democratic organization than the World Bank, where voting rights are purchased. This arrangement gives developing countries a much greater influence over the organization's operations.

The FAO Council

The Council meets yearly, and in effect, serves as the Conference's executive committee. It consequently oversees the detailed administrative efforts of FAO's Director-General, especially during the interim year between Conference meetings (see Phillips 1981, 23). In 1977, at the Nineteenth Session of the FAO Conference, the size of the Council was increased to its present 49 seats. The size of the Council has gradually increased over the years in an effort to match the increase in FAO membership from the original 15 of 1945 (Phillips 1981, 23). The Council's 49 members are divided among FAO's regions. As with the FAO Conference, the political control of FAO and hence FAO Forestry (as with most UN agencies) is in the hands of the poor developing nations while financial control and especially FAO's "extra-budgetary" activities, such as special projects, are controlled by the fewer, but far richer, developed countries. Not unexpectedly, this discrepancy can create serious tensions within the organization over its mission and the allocation of resources to pursue it.

The Committee on Forestry

The Committee on Forestry (COFO) is the Forestry Department's formal link to the FAO Council. COFO was established in 1970 when forestry became a department. It meets every non-Conference year and serves as an advisory group to the FAO Council on matters of forestry. It also oversees the Forestry Department's program of work. At the conclusion of its meeting, COFO submits a report of its discussions and recommendations to the Council for its consideration. Unlike most other FAO Committees, COFO has always been an open committee. Consequently, any member country, upon formally notifying the Director-General, may take part in COFO's work. This makes the Forestry Department all the more a democratic organization compared with other FAO units.

As a result, the political structures that govern FAO and FAO Forestry have tremendous influence over the latter. FAO's program of work, as it is called, must be approved by member countries via the Conference and Council. Since FAO Forestry is only one of many units within FAO, forestry issues must compete with the other pressing concerns of food and agriculture. Historically, forestry has commanded only a small percentage of FAO's Regular Programme funds.[16] This reflects the general lack of attention that forestry issues have received from FAO and its members. It also illustrates the struggle that FAO Forestry officials have had before them in attempting to generate more interest and support. As one FAO Forestry official put it, forestry's call for attention is often a call in the wilderness.

But most important, the policies that emerge from these structures become "marching orders" for the organization. With the exception of COFO, which is only an advisory committee, FAO is dominated by agricultural interests. This representation, along with the inherent greater concern about food over trees, helps to perpetuate the historic disfavor of the forestry sector. In an attempt to overcome this disfavor, FAO Forestry strove to more clearly link forestry to food security with a series of publications in the late 1980s. In sum, although FAO Forestry Department has slowly gained more respect and resources, it has never been a favored unit within FAO.

The United Nations and FAO's Secretariats

The United Nations General Assembly and FAO's Secretariats can shape FAO Forestry's activities dramatically and rapidly through their requests for information. Information is often needed for essential publications or background papers for conferences. FAO Forestry's serious work in fuelwood and charcoal, for example, came in the form of information needed for the 1981 United Nations Conference on New and Renewable Sources of Energy held in Nairobi, Kenya. Likewise, FAO's expanded interest in environmental degradation and monitoring of forest resources was the result of the Stockholm Conference in 1972 and, more recently, the UNCED conference in Rio De Janeiro, Brazil, in 1992. A significant amount of FAO

Forestry's Regular Programme activities are geared towards the need of other UN organizations.

FAO Forestry's Dual Nature: Regular and Field Programmes

FAO Forestry can almost be viewed as two separate organizations. The Regular Programme consists of the Forests Resource Division, Forest Industries Division, and the Policy and Planning Service. The Field Programme is administered by the Operations Service. In addition to their different functions and staff, the programs differ mostly in their sources of funding. The Regular Programme receives its funds from member-country dues to FAO Forestry. Except for occasional political actions made by some wealthy nations (e.g., as when United States under Reagan halted payments of its membership dues to U.N. agencies), this source is generally quite stable, although support more recently has begun to wane again with the present conservative Congress in the United States. The Field Programme operates from variable extrabudgetary sources from UNDP and donor governments that generally provide funding for field projects in specific countries. Although the programs are run almost as separate organizations, each with separate sources of financial support, some program staff overlap. FAO Forestry's Regular Programme staff provides technical support, or "backstopping," for its field projects. This does create some problems for the Regular Programme staff since its own workload can be dramatically affected by its backstopping duties. This certainly has been the case with the growth of the Field Programme under donor funding through TFAP. By 1993, the ratio of Regular Programme to Field Programme staff had increased to 1:6, compared with 1:1.3 in other divisions within FAO (FAO Council 1993).

A look at the location of its employees would seem to indicate that FAO Forestry is a decentralized organization, with most of its staff overseas. In this sense, it is very similar to CARE USA, discussed in the next chapter, although FAO Forestry lacks permanent field missions.[17] Because of the dual organizational structure, to simply classify FAO Forestry as a decentralized organization would be a bit misleading. By design, the Regular Programme is more centralized, with most of its staff stationed in Rome where the staff oversee FAO Forestry's more diplomatic and academic functions. FAO Forestry staff members associated with the Regional Commissions are part of the Regular Programme. Similarly, by design, the Field Programme staff members are mostly in the field, but many are on temporary contracts and are not usually career FAO employees.

FAO Forestry's Task Environment

Identical to that of the World Bank, FAO Forestry's task environment is composed of its member countries and other development organizations. Unlike the other two organizations of this study, however, the FAO's task environment is somewhat more formidable in its influence.

FAO Forestry's dependent character comes from its structural features. It is an agency designed to serve others and works only with the forestry units of its country members. And as a service agency it can never refuse work asked of it. CARE USA, on the other hand, has that option, while the Bank is in a better position to negotiate conditions. FAO Forestry is also only a small unit within a much larger organization, and as an international governmental organization it is expected to be cooperative with other such organizations. These realities greatly reduce FAO Forestry's ability to control its own programs and activities. It is far more reactive than either CARE or the Bank.

Relationships with Donor and Recipient Governments

The member countries of FAO politically control the activities and operations of FAO Forestry Department through the FAO Conference and Council. These structures directly affect FAO Forestry's Regular and Field Programmes.

Donor and recipient countries shape FAO Forestry's field activities in a number of ways. The most obvious is in the form of trust funds (also known as extrabudgetary funds) that donor countries make available on a voluntary basis, usually for specific field projects. These independent projects tend to fall in line with the themes of the Regular Programme, but they may not. Seldom is the money available without "strings" from the donor, generally in the form of who and what are acceptable recipient countries and projects. The donor and recipient countries obviously determine what are mutually acceptable projects. In addition to the obvious control of resources and project ideas, the activities of donor and recipient countries obligate FAO Forestry to play the role of intermediator and, at times, negotiator between the proposed projects and recipients. The project development process, at its extremes, involves "recipient-up" and "donor-down" projects. Recipient-up proposals emerge from recipient countries. FAO Forestry assists by locating a willing donor to support the project. Donor-down projects are those in which FAO attempts to find a country that has a need to fit the type of project the donor country would like to fund. Most proposals, however, fall in between, with both project and funder already in mind. In its role as negotiator, as one FAO Forestry staff stated, FAO essentially is biased in favor of the recipient countries. FAO Forestry will not promote a project it does not consider technically relevant and important to the recipient country. But, as I discuss later, FAO can use donors to set up conditions for performance, as was done in its original FLCD program with Sweden.

Other Development Organizations

FAO Forestry has had a long history of working closely with other development organizations in and outside of the United Nations community. As mentioned earlier, one of FAO Forestry's first major activities was to work with the United Nations Economic Commission for Europe in establishing a vital timber trade agreement with several countries in Eastern Europe to further postwar reconstruction. FAO Forestry has worked with other organizations, such as UNESCO, WMO, UNEP, IUCN, and many others, mostly in a temporary fashion, usually to collect or disseminate information. An important illustration of this was FAO Forestry's collaboration with UNDP, the World Bank, and WRI in the development of and support for TFAP. More recently, FAO Forestry worked closely with UNEP (United Nations Environment Programme), UNESCO, and IUCN (International Union for Conservation of Nature and Natural Resources) on the protection of biological diversity, and with WMO (World Meteorological Organization) and IPCC (Intergovernmental Panel of Climate Change) in preparation for the UNCED Conference (Saouma 1991)

FAO Forestry, through its parent organization, has established several permanent agreements with other autonomous organizations. Two stand out. Each will be elaborated later. The first agreement was reached in 1964 between FAO and the Bank and established the FAO Investment Centre. This unit, housed within the Development Department of FAO, attempts to identify appropriate and technically sound investment opportunities for eventual Bank funding. The second agreement is with UNDP for its financial support of most of FAO Forestry's field projects.

Financing FAO Forestry

Unlike the World Bank, which generates its own resources by creating loan agreements, FAO Forestry is totally dependent upon the resources of others. FAO Forestry's Regular Programme is funded by dues from member nations. The FAO Conference approves the allocation of funds to specific departments and programs. In an earlier monograph (Brechin 1989, 174), I showed that although funding for FAO Forestry had increased dramatically from 1958 to 1987, financial support for forestry in percentage of Regular Programme expenditures has decreased over the years. This trend has continued (FAO Council 1990, 1993).

FAO Forestry's Field Programme is financed by other sources: UNDP, trust funds from member countries, and the Technical Cooperation Programme (TCP), FAO's own internal source. With 49 percent of FAO Forestry's field projects supported by UNDP during the period 1980–90, it is the largest single source for FAO Forestry's field projects. Trust funds from donor nations grew steadily from $18.7 million in 1980 to $34.4 million in 1990 (both figures adjusted to 1992 dollars) and have become a key source of support for field projects (Brechin 1994). Trust funds exceeded UNDP funding for the first time in late 1988, largely as a result of coordinating trust fund giving through TFAP (see Muthoo 1991). It has been the trust funds, not UNDP, that have been the most supportive of FAO Forestry's community forestry fieldwork. FAO's dependence on UNDP as the largest source of field project revenue has greatly limited FAO's work in this new activity.

TCP was created in 1976 and is derived from internal FAO funds. TCP support is in the form of small grants for projects, generally to meet urgent or unprogrammed needs for technical and emergency assistance. During the period 1980–90, FAO Forestry sponsored 197 TCP projects with an overall budget of $23.9 million (1992 dollars) (see Brechin 1994). TCP, because of its limited size and its emergency criteria, will never become a major source of funds for FAO Forestry field projects. It does, however, serve an important role in developing new projects and in providing bridging support for the projects between funding.

FAO Forestry's Core Technology: Professional Knowledge for Governments

Unlike the World Bank, FAO Forestry is not a funding agency. Outside of TCP, FAO does not provide grants or loans. "What we have to offer," one FAO high-ranking Forestry official told me while reflecting on FAO's role, "is professional knowledge." Not surprisingly, forestry knowledge at FAO has been centered around the professional training and international experience of its staff.

FAO Forestry's core technology of professional technical knowledge has always been grounded in its Regular and Field Programmes. This basic division of labor has also always been augmented by several overlapping conceptual or thematic activities. These themes, or what FAO Forestry has called its "Technical Programmes," overlap more or less with the department's organizational structure.[18] These programs have changed with time. During the primary period of this study, 1980–90, there were four Technical Programmes: Forest Resources and the Environment, Forest Industries and Trade, Forestry Investment and Institutions, and Forestry for Rural Development (which included community forestry). Under the Technical Programmes, the headquarters and field components can be folded together to become more complete and integrated technical packages for member nations. These programs may be implemented across organizational units. This has become true of the Rural Development Programme. FAO Forestry, through its Regular, Field, and Technical Programmes, provides training, education, expert advice, and development assistance to its member countries. In the process of providing those services, FAO serves as neutral forum for discussions on international and regional forestry matters; and, consequently, by its work, position, and legitimacy, it is as an international leader in forestry.

As noted earlier, an important part of the Regular Programme's activities centers on the collection and dissemination of information. One Regular Programme official referred to himself and his colleagues as "brokers" who collect, trade, and dispense knowledge. The idea of FAO Forestry's Regular Programme as a brokerage house is illustrative. There is also an academic quality to FAO Forestry's work. The headquarters' staff consists of educators and trainers as well as diplomats who strive to increase their understanding of the complexities of international forestry at the regional and global level. They look for patterns among problems and solutions to forestry matters so that they may pass along that information to their member-country clients.

FAO Forestry also dispenses its professional knowledge to meet the specific needs of country members through its Field Programme. This activity reflects more the development agency quality of FAO Forestry. Here, in sharp contrast to the Bank, it is an *executing agency* for projects sponsored by funders. The FAO Forestry Field Programme is highly varied. Projects may range from a one-month visit by an expert to advise a government on policy or planning issues, to multiyear, multimillion-dollar budgeted projects with large international and domestic staffs. Many field projects contain training components. This is largely the result of FAO Forestry's role as a service agency to government forestry agencies, where "educating" government forestry personnel is often the primary objective of the project. This is important to note, since many of FAO Forestry's projects, or large parts of them, are less geared to the achievement of actual technical objectives (such as planting a certain number of trees) than they are to the general training of government personnel in tree planting and forest management. This is an important trait of FAO that can affect the performance criteria of its projects, which is discussed more in Chapters 4 and 5. FAO Forestry's experiences with field projects, however, provides not only dividends to the organization but also a worldwide wealth of empirically based knowledge.

A Shift in FAO Forestry's Technology

This is a key question in this discussion of FAO Forestry's technology. What exactly constitutes the professional technical knowledge of forestry? Historically, forestry has been highly technical, narrowly focused on the management of trees (as in plantations), and natural forests primarily for large-scale industrial and commercial purposes. Knowledge of forest products has been similarly technically specific. Fortmann (1986) called this kind of forestry large-scale production of fiber for commercial use. This interpretation of the technical content and aims of forestry, as traditionally conceived and practiced, is generally accepted and agreed upon. When reviewing FAO documents, especially before the mid-1970s, forestry is described in no other terms (see Brechin 1989, 179 n. 36). Consequently, until the advent of community forestry, for all practical purposes forestry meant management of trees and forests for industrial or conservation (e.g., watershed protection) purposes, or what I simply call "traditional forestry" (de Montalembert 1991). As an example, one of the general sources on FAO noted that the pulp and paper industry had during this era the longest and deepest involvement at FAO Forestry (FAO 1985a; 59).

The formal creation of the Special Programme for Forestry for Local Community Development (FLCD) in 1979 marked an important moment in its history. That occasion and the institutionalization in 1980 of the Technical Programme called Forestry for Rural Development officially signaled institutional acceptance of the concept of community and rural development forestry, as FAO Forestry formally incorporated the programme into its agenda. In doing so, FAO Forestry also altered significantly its core technology and character. It had broadened its technical concept of forestry to include a wide range of activities involving the use of trees by rural people and communities, not just by national governments and industries for the production and export of timber. As I have described in some detail elsewhere (Brechin 1989, 20–23), community forestry embodies a different philosophy and set of technical concepts (see also Westoby 1987, 1989).

It is important to note that community forestry has not replaced traditional forestry within FAO. Traditional forest management and industrial forestry are alive and well. There is still tremendous demand for that type of knowledge and expertise. It is perhaps more surprising that community forestry and traditional forestry are not simply parallel programs. The aims of community and rural development forestry have partially seeped into the traditional forestry units and have shifted the focus to the local level and to people and their work, as in sawmill operations (Brechin 1989, 180).

Organizational change, especially change in an organization's technology, does not occur easily. Thus, it is essential that we explore in greater detail how FAO became involved in community forestry. As we shall discuss later, the more dependent the organization is on outside resources the easier it is for the organization to alter its core technology.

An Assessment of FAO Forestry's Core Technology

Fontaine (1985, 13–14) suggested that there had been an important qualitative change in the application of FAO Forestry's professional knowledge. In particular, he suggested that in more recent decades, FAO Forestry had moved away from the international focus of its beginnings to one centered more on the problems and issues of individual nations. In short, Fontaine suggested that the Field Programme activity had evolved to a point where it had become detrimental to the Regular Programme. Although critical of this change, he did not elaborate in any detail. It seemed Fontaine felt that the emphasis on individual nations had detracted from the efforts of international organizations to promote interstate relationships and the development of a world economy. He seemed to imply that field projects have provided FAO an exceptional source of experience and information but also a refuge from international tensions. If this is the case, then Fontaine, wittingly or unwittingly, supports the perspective best expressed by the internationalistist David Mitrany that world peace can best be accomplished through the development of interstate relationships enmeshed in mutually beneficial economic relationships, thereby making war a more foolish undertaking. It may also reflect the decline of a more hegemonic era when the international agenda for forestry was exclusively determined by the Western industrialized countries.

In looking at budgets as a measure of dominance, the figures support Fontaine's argument that the Field Programme has overshadowed the activities of FAO Forestry's Regular Programme.[19] It also must be realized that the country members of FAO set the tone for FAO's activities. The Field Programmes have grown because they are desired by members. I would agree with Fontaine, however, that country field projects are organizational tasks that are not particularly political. It is indeed an arena where the organizations can grow and prosper with little political resistance. Yet, while this may be true, FAO Forestry's core technology has been vastly shaped by the needs of its task and broader institutional environments over the resistance from traditional foresters.

Strengths and Weaknesses of Professional Knowledge

The only aces FAO Forestry can play are its professional technical knowledge, along with its international legitimacy in forestry matters. As we have seen, FAO Forestry, as an organization, is greatly constrained. Its Field Programme is particularly vulnerable especially because it is funded primarily through donations provided on a project-to-project basis.[20]

Consequently, FAO Forestry is limited by its inability to more freely pursue its own visions and activities, since the power of the purse lies in the hands of others. Likewise, it does not have the power to make other organizations perform as it wishes. In addition, it is also limited by the constitutional controls that restrict is ability to act. As already discussed, FAO's programs and activities are controlled directly by its members through the FAO Conference and Council.

While this particular imbalance of power is certainly desirable from the member countries' standpoint, it does place FAO Forestry in delicate and contradictory situations. As both a professional service organization and the world's forestry agency, it must be a follower and leader at the same time. A common remark by officials at FAO Forestry was that FAO must serve its members, not lead them by the hand. Consequently, pronouncements of directions and policies without explicit approval of its members would be a serious breach of protocol. In its activities and programs, FAO can only be where its members are in recognizing their own needs.

Being leaders and followers at the same time, however, places FAO Forestry officials on a management tightrope. This tends to make FAO Forestry a slow and cumbersome organization, as new ideas and directions need to be discussed and consensus achieved at various political levels before formal action can begin. Such action may take years to complete. These reasons are precisely why FAO Forestry was a bit slower than the Bank *in formally* establishing structures related to community forestry and forestry for rural development.[21] By design, it is very difficult for FAO Forestry (as well as other UN agencies) to take bold and decisive positions and actions on international matters. Consequently, it is absurd to criticize FAO's lack of leadership too harshly. FAO and most other international organizations were designed to respect national sovereignty and first achieve consensus before acting, thereby reducing considerably its ability to actively lead.

The Politics of Meetings and Conferences: The Role of Diplomats

FAO and FAO Forestry attempt to manage this delicate balancing act with the aid of several mechanisms, such as the art of diplomacy, the power of their professional knowledge, and the legitimizing functions of international conferences. Armed with extensive information and respected technical expertise, FAO Forestry has the ability to raise issues and help set the agendas at conferences and meetings. Through the art of conference management and diplomatic persuasion, FAO officials can help guide members, gently, to where they think the members should be in programmatic interest. Given FAO's role as a servant to member governments, it must remain a quiet leader. Consequently, its officials must be *diplomats* first and *foresters* second. Professional expertise must take a back seat to political necessities.

In addition to developing a consensus for action, these meetings, especially those of advisory groups and expert consultations, keep the FAO staff in contact with leading authorities on a range of forestry matters. This allows staff members to keep abreast of the most recent and accurate information, thereby enhancing their professional knowledge.

The Politics of Publications

Perhaps one of the most striking aspects of FAO is its long list of technical publications. Publishing useful and often important documents has long been an essential part of the character and technology of FAO Forestry. These publications serve a number of functions for the organization. As noted by one staff member, FAO's publications strongly reflect the efforts of the organization.

Many of these publications are state-of-the-art reports about a particular subject. They often reflect the collective experiences of a number of member countries that have similar problems or have achieved success at solving them. Consequently, the documents become important tools for the training and education of member-country foresters.

Not only do these reports compile useful case study materials, they tend to include FAO's own professional recommendations. FAO publications thus record the organization's professional judgments on a wide range of forestry subjects. These publications institutionalize organizational knowledge, but perhaps more interestingly they become neat summaries in which the organization can promote its own competence and value.

With the notable exception of *Unasylva,* which itself is in large part a public relations tool, FAO Forestry publications are not the product of the collective organization but of its professional subunits, acting somewhat independently of one another. Publishing has become an important mechanism by which these subunits can stimulate demand for their expertise and enhance their status within the larger organization. The documents can also play a vital role in "sales pitches" to interested donor countries. Publications provide tangible proof of FAO's expertise on a particular topic, useful in demonstrating competence to recipient and donor countries alike. As one long-time FAO Forestry staff member remarked, "It is important for FAO to appear to be on top of the issues."

My critique of FAO Forestry's technology strongly points to the reality that its professional forestry knowledge is its major strength. It is nearly powerless in any other sense. It relies wholly on its ability to articulate technical positions and rationale and on its diplomatic skills to promote an appropriate course of action. It can only advise and recommend, not command. Without seriously violating diplomatic protocol, it cannot withdraw its services from members it may disapprove of. It cannot combat or call specific attention to member-country problems such as corrupted officials or unfortunate events that are endemic to political situation or cultural heritage (see Brechin 1989, 186). Although FAO Forestry may, through diplomatic manipulation, nudge member countries in one direction or other, the bottom line remains that it is constrained to serve its members as they wish to be served. This is much more the case than with either the World Bank or CARE USA. The Bank has important powers since it controls or at least moderates vast financial resources that others want and need. If CARE is dissatisfied with government performance or terms of agreement, it simply leaves or refuses to work in that country. FAO Forestry has neither the luxury of capital nor the capacity to exit.

Summary of FAO Forestry's Character

As an organization, FAO Forestry is seriously constrained. It is not a funding organization that gives or loans money to others. It operates solely on resources given to it by others. As a service agency, principally of developing countries, it attempts to meet the expressed needs of its clients. Its members must approve of its actions, which are determined by consensus through the diplomacy of meetings, conferences, and publications. The agency's professional technical knowledge and its legitimacy to act as a world leader on matters of forestry are its greatest strengths.

In spite of its constrained nature, FAO Forestry performs a number of useful activities. It provides forums for discussion of forestry issues; is a classroom were information and new ideas are shared; is a thinking organization that nurtures the forestry intellect; evaluates the state of international forestry; reflects on trends, both past and future; acts as a politically neutral executing agency for forestry projects; and advises governments in matters of forestry policy and planning. Finally, the organization records and houses, as a permanent record, the activities and events that compose the field of international forestry.

The work of FAO Forestry, as we have already indicated, is dependent upon three very different funding sources. Its fieldwork has depended mostly upon funding from UNDP and grants from individual countries, while its work at headquarters is supported largely by membership dues. This gives the organization a dual character, with the Field Programme (in budget and personnel) overshadowing the Regular Programme. The Field Programme activities also depend more upon the wishes of donors, with most projects geared to training or working with host government staff. The somewhat less-dependent nature of the Regular Programme, and its role in collecting and disseminating information, allows it to work more at the "cutting edge" of international forestry issues. Of our three organizations, FAO Forestry is the most constrained, but also the most "academic" and diplomatic.

Since the 1960s, the activities of FAO Forestry, along with most UN agencies, have shifted from broader international issues and concerns to working more directly with individual countries on development matters. It has shifted as well from the development of the forestry sector to demonstrating the value of forestry in the process of development, with a recent emphasis on rural areas and integrating forestry into the activities of rural people.

Chapter 3

Helping Poor Communities Plant Trees: CARE USA

CARE USA was founded in November 27, 1945, when 22 American organizations joined together to channel aid to European survivors of World War II. Today, it is the largest nonsectarian, nongovernmental, nonprofit development and relief organization in the world. In fiscal year 1993, CARE alone spent over $405 million on its projects and services (CARE 1993). As of its fortieth anniversary in 1985, CARE had provided goods and services to people overseas valued at over $4 billion (CARE 1986a, 2).

CARE is a decentralized organization with established field missions in developing countries. CARE's decentralized structure has had enormous influence on its community development programming core technology by encouraging mission-centered decision-making and -planning processes. Although decentralized, CARE's work in the field is often considered by many as somewhat "top-down" in approach, at least compared with many other private voluntary organizations (PVOs). It is certainly more "bottom-up" and participatory than the World Bank, and at times more so than FAO. CARE maintains significant control over its task environment through its status as a private humanitarian organization and through the generation of its own financial support. Most of CARE's financial support comes from government-sponsored surplus food and contract grants. Its internal source of funding largely comes from public fund-raising through effective promotion of its overseas relief and development work in the world's poorest countries. CARE's internal funds give it considerable freedom and flexibility in its efforts.

For years, CARE has enjoyed a fine reputation among the general public, governments, and other development agencies (Campbell 1990; Cazier 1964, 445; Katz 1974, 185). Throughout its history, CARE has been considered a flexible, people-to-people organization that has consistently provided appropriate inputs in a timely fashion (Campbell 1990; Cazier 1964, 442; Katz 1974, 180). Since 1974 CARE has considered itself, with some justification, a leader and innovator in the field of agroforestry and natural resource management in general (Campbell 1990; CARE 1987b). As we shall see in later chapters, CARE's work in community and rural development forestry, compared with the Bank and FAO, is both more and less impressive in several ways.

Of the three organizations, CARE is the only one that can design, finance, and implement its own projects. Like FAO, but unlike the Bank, CARE's projects are financed as grants, not loans. It is also the only one of the three that works strictly at the community level, with its programs built around the concepts of self-help, decentralized decision making and participatory development (although, as was just mentioned, other PVOs are often better at participation than CARE is). CARE's program in forestry is for the most part geared toward local communities.

The Development of Community Forestry at CARE

CARE's systematic involvement in forestry dates back to the mid-1970s, and in agriculture, to 1950 when it expanded its activities by including seeds and plows in refurbished CARE packages. Additional activities, such as rangeland development, joined agriculture and forestry as part of CARE's Agricultural and Natural Resources (ANR) unit as it continued to expand its work in the natural resource management sector. From 1981 to 1987, forestry projects grew from 0.2 percent to 2.8 percent of CARE's total field program (see Brechin 1989, 267), and rose to 4.5 percent in 1992 (Brechin 1994; CARE 1992).

The Character of CARE's Forestry Program

CARE always implements its own forestry projects. In addition, it may also design or fund them as well. Essentially, all of CARE's forestry projects are either on-the-ground applications with specific communities and villagers as target populations, or training programs for government staff to better assist individuals and communities. CARE's operating procedures call for government agencies such as forest services or similar units to work with it in implementing the projects. CARE's degree of direct involvement in project implementation tends to be inversely correlated to the degree of competence and abilities of CARE's complementing agency. As was observed in Brechin and West (1982), with its presence and active involvement in the projects, CARE can provide timely "bench" support to the government's efforts by being ready to step in and provide the assistance needed to make the project successful. CARE may also run training programs for government personnel. But unlike FAO Forestry, CARE's primary target population remains the rural poor and not the government agency staff.

Processes of Institutionalization: CARE's Beginnings in Community Forestry

Although, like the Bank and FAO, CARE's work in forestry has been influenced by the changes found in the larger institutional environment, it seems to have been much less directly affected by them. CARE's community forestry work was more directly the result of attempting to fulfill a "bottom-up" need that itself could be understood and identified precisely because of the larger paradigm shifts to rural development and environmental management (see chap. 1 and 2). Unlike the Bank and FAO that were more directly involved in pursuing and reflecting the institutional change, CARE's beginnings seem much more genuine.

The most striking aspect of CARE's involvement in forestry is that it was initiated from community-level field sites. Almost simultaneously, CARE became involved in forestry projects in two different countries and continents, one in Niger and a second in Guatemala. In both cases, CARE was approached by field personnel of recipient governments.

CARE's first forestry activity began in 1974 when a Nigerian forester and his Peace Corps assistant approached the CARE Mission in Niamey, Niger, with an innovative agroforestry project to establish windbreaks along the floor of the Majjia Valley, a relatively rich agricultural area of central Niger. The Majjia was suffering severe ecological damage from a combination of the valley's comparatively high population density and the Sahelian drought of the early 1970s. The area's future agricultural productivity was in serious jeopardy (Delehanty, Hoskins, and Thomson 1985; Leach and Mearns 1988).

CARE accepted the proposal; and with its financial support, the foresters, by December, 1974, were able to establish a primitive nursery in the village of Karaye.[1] Planting of seedlings for the windbreak (and live fences) began during the rainy season of 1975 (Delehanty, Hoskins, and Thomson 1985, 2). The now-famous Majjia Valley Windbreak Project ended in 1992 after 19 years of support by CARE and other donors. At the end of CARE's direct support, about 840 kilometers of windbreaks had been planted, protecting around 11,000 hectares of agricultural land (ANR 1994).

CARE's second forestry project began later in 1974 when the CARE mission in Guatemala was asked to provide small quantities of Food for Work (FFW)[2] provisions to a number of selected existing community forestry projects sponsored by the government. CARE became more intensely involved in the following years, starting in 1975 when the mission provided equipment in addition to FFW to project pilot sites (Chemonics International 1983).

The forestry projects were originally the suggestion of an Organization of American States study team that recommended immediate action to correct the severe erosion problems of the highlands of Guatemala's Altiplano and Oriente regions. The projects, which integrated soil conservation measures with tree planting on mountainous slopes, sought to improve conditions for small farmers. The projects were organized at the community level and were jointly managed by the Instituto Nacional Forestal (Guatemala's forestry department), CARE/Guatemala, and the Peace Corps/Guatemala (Chemonics International 1983). With CARE's participation and financial support, the number of project sites expanded from 9 to 50 between 1975 and 1978 (Chemonics International 1983, 9). From 1982 to 1985, over 11 million seedlings were planted from 60 community nurseries (CARE 1985c). CARE's forestry program in Guatemala still continues as of my writing this book.

Although CARE missions have been planting trees systemically since the mid-1970s, CARE's formal program in forestry was first established when it received a multimillion-dollar Renewable Natural Resources Matching Grant from the U.S. Agency for International Development in 1981. This grant created a Renewable Natural Resources Coordinator's position at CARE's New York headquarters and provided umbrella funding for 10 existing and new forestry CARE projects. As a result, CARE's work in this sector grew rapidly. In 1986, an ANR unit was created and housed in CARE's Program Department to better coordinate the effort. By the late 1980s CARE had over two dozen community forestry projects operating in nearly 20 countries on three continents. In 1994, CARE could boast of 90 forestry projects in 28 countries (ANR 1994). The tremendous support of CARE's efforts came mostly directly from USAID and the general public, but was ultimately the result of cultural shifts from the 1960s and 1970s that supported the need for sound development strategies involving better management of the environment.

The History and Character of CARE USA

This section will focus on the historical forces that have shaped CARE's ability to work directly with local communities, its shift from Europe to the Third World, and its movement away from relief packages to integrated community development programs. Reviewed as well are the pressures placed upon it to remain both efficient and effective, and its international growth of financial support.

As with the World Bank and FAO, CARE was a product of World War II. CARE, which originally was an acronym for Cooperative American Remittances to Europe (Campbell 1990; Cazier 1964, 34), was created as a private voluntary aid organization when 22 private American organizations, many of them religious, joined together to more efficiently distribute badly needed essentials to the peoples of Europe. CARE was also in part a product of the U.S. government's desire to coordinate voluntary relief to Europe.

From Chaos to Cooperation in Voluntary Relief

During the war, hundreds of PVOs were simultaneously collecting money from the American public and sending relief supplies to Europe. From 1939 to 1942, some $100 million had been raised by these organizations (Cazier 1964, 4). Unfortunately, with the absence of any regulation or coordination of their activities, there was much confusion, duplication of effort, fraud, and, consequently, waste of valuable resources. It was also difficult for the U.S. government to be sure that private groups were not supplying the enemy (see Cazier 1964, 4).

In 1941, President Roosevelt established a commission to study the problem of voluntary relief. It recommended the creation of a federal agency to regulate the voluntary agencies. As a result, the War Relief Control Board was created by the government in July 1942 to "establish order out of war relief chaos" (Cazier 1964, 5). The Board, provided with extensive powers, demanded greater cooperation and coordination among the agencies and began to seriously regulate their activities. Only 90 voluntary agencies of the hundreds survived the Board's strict regulatory requirements (Cazier 1964, 5). Efficiency, effectiveness, and especially cooperation became the watchwords of voluntary relief. The concept of CARE emerged from these events.

As a result of the Board's action, a number of new cooperative organizations were established. One was the National War Fund, whose quick success showed that coordinated fund-raising was an effective way of tapping donations from the public. Another success was The United Jewish Appeal, which had been in operation since 1939 (Cazier 1964, 6). The American Council of Voluntary Agencies for Foreign Service (American Council) was a cooperative organization that was created as a result of government regulation and that later had an important impact on CARE. It was established in October 1943 as an umbrella service organization for other voluntary organizations involved in overseas relief and reconstruction (Cazier 1964, 7; see also Reiss 1985).

CARE as an organization can be traced to Arthur Ringland, who served on the President's War Relief Control Board (Campbell 1990; Cazier 1964, 5, 29; *Washington Post* 1981). Following World War I, Ringland was an administrator for the American Relief Administration Warehouse (ARAW) program. ARAW had directed a component of the U.S. relief efforts in Central Europe following World War I. ARAW provided an effective way for private citizens to contribute to the relief of Europe. With the ARAW, the U.S. government had created a unique mechanism for the country's large immigrant population to help people "back home" through the use of remittances and packages. Under this program, relatives and friends in the U.S. could purchase drafts or remittances that were then sent to loved ones overseas. The remittances were presented to program personnel at specified warehouses in European countries, and the packages were released (see Campbell 1990; Cazier 1964, 26–29). Ringland, as consultant to the War Relief Control Board, suggested reviving the people-to-people package concept.

Ringland was able to stir considerable interest in the idea. The United Nations Relief and Rehabilitation Administration (UNRRA), created in November 1943, provided one of their consultants, Lincoln Clark, to help promote the package concept (Campbell 1990; Cazier 1964, 7, 31). In addition, other organizations such as the American Red Cross, the National War Fund, and the American Council supported the idea. Clark contacted Wallace Campbell, Secretary of the International Committee of the Cooperative League of the U.S.A. (Campbell 1990; Cazier 1964, 31). The Cooperative League had been formed before the war to support the development of cooperatives in Europe. Together, Clark and Campbell sketched out a plan for a PVO that would have a broad base of support in the United States and the ability to distribute packaged relief supplies overseas.

The American Council was very receptive to Campbell's inquiries. Many of its members did not have counterparts overseas. Hence it was easier for the members to raise money in the United States than it was to distribute the actual relief supplies overseas (Cazier 1964, 33). As a result, many of the

American Council members needed an organization to represent them and distribute their supplies in Europe. With Campbell's Cooperative League members in Europe providing the overseas outlet points, the remittance-package concept suggested by Ringland, and perfected by Clark, and the broad-based U.S. fund-raising support from 22 voluntary organizations from the American Council's membership, the critical ingredients of CARE were in place. It was incorporated on November 27, 1945.

Throughout its early history, CARE encountered numerous difficulties, some which threatened its very existence. It was, however, from the very beginning, appreciated and directly supported by the U.S. government. It fit nicely the requirements of the War Relief Control Board's mandate. It proved to be a very effective, efficient organization. Most important, CARE maintained broad-based support as a cooperative of a diverse set of relief organizations. The War Relief Control Board fully sanctioned and fostered CARE's activities. Even UNRRA facilitated CARE's development. UNRRA shared with CARE its options on U.S. Army surplus 10–1 rations, which provided CARE with its first supplies (Cazier 1964, 40). CARE reorganized and supplemented the 10–1 rations to fit family needs, and the first of the now-famous CARE packages was on its way to Europe. The first CARE package arrived in Le Havre, France, on May 11, 1946 (CARE 1986a, 6; Cazier 1964a, 58; Katz 1974, 179).

CARE Packages to Europe

CARE's relief to Europe was patterned after the ARAW package program used at the conclusion of World War I. CARE, however, made several improvements that added more personal touches for the public. First, CARE offered a guarantee to its donors. If a remittance was not received by a loved one in 120 days, the donation was refunded. To prove that a package arrived, CARE returned to each donor a copy of the signed receipt. As noted by Cazier (1964, 59), these changes proved popular. It also made CARE even more responsible to its contributors.

In September 1946, CARE's board hired Paul French as its new executive director. This was a turning point for CARE in its young history because French orchestrated a number of important changes (Cazier 1964, 73). First, he drastically reduced the size of CARE's staff, which was considered by many as too large to match the volume of its work. More important, French moved CARE away from paid advertisement. He was able to solicit the support of the U.S. Advertising Council (Campbell 1990). With the Advertising Council's full support, which it received over the next 10 years,

CARE received the equivalent of $8 million annually in free advertisement (Cazier 1964, 73). This exposure helped make CARE a household name.

To promote the CARE story throughout the United States and to tailor its appeal to local interests, CARE established field offices in every major city (Campbell 1990). By the end of 1947, it had over 400 office outlets and a growing list of donors (Cazier 1964, 74). To increase efficiency, CARE revamped its internal office procedures at headquarters and consolidated its overseas operations. To help spur sales, CARE lowered its basic package price from the original $15 to $10. French also encouraged the development of new CARE packages and the redesign of existing ones to meet a wider range of needs in Europe. By the spring of 1947 CARE finally achieved financial soundness, and by year's end had already become the second most active voluntary relief organization in the U.S., with sales of 2,608,116 packages valued at over $25 million (Cazier 1964, 84–85).

With the changes administered by French, CARE demonstrated the efficiency, flexibility, and ingenuity that was to become an important part of its character. As noted by Cazier (1964), "CARE had established an enviable reputation as a result of its relief work. It had proven to be flexible in response to human need, deft in the business of relief, and imaginative in enlisting donor support" (86).

The Expansion of CARE

As it strove to achieve financial stability, two additional problems plagued CARE in its early years. The first was the issue of designated versus undesignated relief. The second was the expansion of CARE's work outside of Europe. The solution of these two closely linked problems changed the organization's fundamental character.

CARE's original mandate was to "supply or make available goods and services to needy persons in foreign countries . . . for the purposes of relief, rehabilitation and reconstruction" (Articles of Incorporation 1945; see also Cazier 1964, 398). In spite of its early work in Europe, CARE was not limited at its inception to serving only the immediate relief needs of Europe. In fact, one of its founding fathers, Wallace Campbell, urged the CARE board as early as 1947 to broaden its work to include rehabilitation as well as relief projects (Campbell 1990; Cazier 1964, 398). Given the realities and immediacies of the war's devastation, CARE's principal focus in its formative years remained the provision of emergency relief supplies to needy Europeans. This proved to be a curse as well as a blessing. The need of relief for Europe was great, and there was tremendous support for it. Such a mission also created the image to many that Europe was CARE's only focus. Many donors as well as some of CARE's board members thought CARE's mission was in fact limited only to *relief* for *Europe* through *designated* packages (Campbell 1990).

In his history of CARE, Cazier frequently mentions the heated debates among board members over the nature of CARE's operations. Because of CARE's efficiency and effectiveness, there was considerable pressure externally as well as internally to expand the nature and locations of its operations (Cazier 1964, 197). It was determined very early on that CARE could easily expand its operations in Europe by focusing on general, or undesignated, relief in addition to its more constraining system of designated packages. CARE had, in fact, never refused money from people generally wishing to help the needy of Europe. The board did, however, specifically limit CARE's promotion of that service. The original focus on designated relief did not allow the organization to deliver food or clothing in bulk to institutions such as schools and hospitals.

During the late 1940s and early 1950s there was also a growing awareness by the U.S. government and the private relief agencies of the vast need for rehabilitation in countries outside of Europe (Campbell 1990). Some CARE board members had considerable interest in expanding CARE's work to other regions. Others CARE board members, from other relief organizations, fearing an expanded CARE as a competitive threat to their own organizations, strongly rejected any change to CARE's program of designated relief to Europe. Several members repeatedly suggested the liquidation of CARE after it finished its work there (Cazier 1964, 286). After much turmoil, the CARE board on June 2, 1952, approved the unrestrained expansion of CARE's activities (Cazier 1964, 286). This approval gave CARE the general go-ahead to expand its operations at will. The Board vote for expansion was followed by the resignation of several of CARE's largest religious organizations and original members (see Campbell 1990, Cazier 1964; Katz 1974).

After this change in direction, CARE quickly established missions in all corners of the world. In the early 1950s, CARE placed missions in Mexico, Peru, Panama, Bolivia, Haiti, Brazil, Chile, Honduras, Columbia, Paraguay, Israel, Yugoslavia, Laos, Vietnam, Hong Kong, and Egypt. By the end of 1954, it was operating in 32 countries (Cazier 1964, 243). As a result of its expansion, CARE needed to change its name. In October 1952 CARE became the Cooperative for American Relief Everywhere (Cazier 1964, 242).

CARE again showed its flexibility and responsiveness in providing for the varied needs of people found in these regions. It offered a large assortment of relief supplies and new services. Over the next few years, it created a number of new packages. In addition to food, CARE sent packages that contained textiles, books, plows, tools of all sorts, teacher's kits, and even commercial fishing supplies. New services included the delivery of bulk foods for school lunches or meals at medical institutions as well as technical expertise and equipment for medical care and for the construction of roads, buildings, water systems, and the like.

CARE continued to expand its operations in the health sector when the voluntary overseas medical organization known as MEDICO joined CARE in 1962. MEDICO was a group of health professionals that had been founded in 1958 by Drs. Thomas A. Dooley and Peter D. Comanduras (Kaufman 1971, 68–69). MEDICO's affiliation allowed both organizations to strengthen their work in the health sector. It provided CARE with a sophisticated and dedicated group of nurses and doctors. The union also relieved MEDICO of the increasingly burdensome administrative duties that were engulfing that small group.

Another essential element in the expansion of CARE (and other voluntary relief agencies) was access to the U.S. government's enormous supply of surplus agricultural products and the subsidization of its freight transportation overseas (Campbell 1990). Direct governmental support of overseas relief began as early as 1948 with Title I of Public Law 472 (the Economic Cooperation Act), which provided payment of ocean freight charges for Europe-bound relief shipments of registered voluntary agencies (Cazier 1964, 190). Access to agricultural surplus by private relief agencies was first formalized with the passage of the Agricultural Act of 1949 (Cazier 1964, 342). The components of these two acts were united, enlarged, and refined by Public Law 480 in 1954 and the Agricultural Act of 1956. Under these laws, voluntary agencies such as CARE were given greater access to the agricultural surplus, with the government paying both ocean freight and inland transportation costs (Cazier 1964, 353, 381).[3] As a result, a new era of voluntary overseas relief and development had begun.

Access to agricultural surplus and its free transportation overseas proved crucial for CARE. But as both its program and theater of operations expanded, donations plummeted in 1950 to only a third of the amount in 1948 (see Cazier 1964, 272). For the general public, the crisis in Europe was over and the concept of Third World rehabilitation was yet to become fixed in their minds.

The distribution of heavily subsidized agricultural surplus helped CARE survive its critical transition from a simple relief agency operating principally in Europe to an organization providing more varied and sophisticated relief and development assistance to the developing countries in the Southern Hemisphere. It also fused a very close relationship with the U.S. government that still exists today. By the early 1960s, CARE was again on solid footing, expanding its program to Africa and experimenting more fully in self-help community development projects (Cazier 1964; Kaufman 1971). The present-day CARE began to take shape.

CARE's Self-Help Development Programs

By 1966, the famous CARE package was being phased out in favor of feeding programs for schools and hospitals (made possible by surplus agricultural products) and in favor of more sophisticated self-help development projects (CARE 1986b, 6). CARE's feeding programs grew rapidly. By 1974, through its missions, CARE was feeding 20 million people each day, mostly children, pregnant women, and nursing mothers (CARE 1974). But as CARE became more interested in and sophisticated about rehabilitation, it was clear that packages and even feeding programs were not in themselves sufficient to stimulate long-term development. This interest in sustainable and permanent development pushed CARE in the direction of self-help projects (Campbell 1990; CARE 1986b, 6; Cazier 1964, 397).

CARE's beginnings in self-help projects can be traced to 1950 when it included simple steel plows and other farm tools in refitted CARE packages sent to India, Pakistan, and Greece (CARE 1986b, 6; Cazier 1964, 408; Katz 1974, 180). Books and other sources of technical information were sent overseas as early as 1949 to Finland and later to Germany in cooperation with the U.S. State Department. Even with the general increase in self-help projects, by 1961 such projects were still less than 5 percent of CARE's total overseas program expenditures (Cazier 1964, 443).

It quickly became apparent to CARE personnel, however, that just sending books and crude farming implements would not be sufficient to promote development. In lands of great poverty and illiteracy, skilled people were needed to help teach people how to read and to use a plow. As a result, technical assistance with comprehensive development projects soon became an integral part of CARE's self-help field programs (see Brechin 1989, 233; Cazier 1964, 414–415)

CARE's movement into "total" self-help development required it to increase its flexibility, innovativeness, and responsiveness. To find new work, CARE had to observe its surroundings more thoroughly in order to provide appropriate services or supplies needed for sustained development. CARE also learned very quickly that one organization alone could not create "total" development.

One of CARE's earliest and most successful community development programs was in Mexico in the 1950s (Cazier 1964, 422). CARE's work there laid the pattern for its program in community development for the next several decades. Before it began its operations in Mexico, CARE reviewed the work of other voluntary and international organizations. Instead of simply initiating new projects, CARE also attempted to supplement the existing projects of other organizations, including those of the Mexican Government (Cazier 1964, 422). In each case, CARE thoroughly investigated each project before any commitments were made. In one project, CARE provided essential equipment and supplies, such as books, woodworking and hand tools, sewing machines, plows, midwifery supplies, and recreation kits, to traveling "Mission Teams" that were sponsored by the Mexican Ministry of Education (Cazier 1964, 423). These teams would often construct community centers in remote rural villages to provide facilities in which they and others could conduct educational classes.

One of the most important and innovative CARE projects in Mexico was well drilling. Contaminated water had long been a source of serious health problems in Mexican villages. CARE purchased a drilling rig that was operated by volunteers and supervised by the Mexican Department of Hydraulic Resources (Cazier 1964, 425). Abundant, potable water went a long way toward improving life in Mexican villages where CARE worked.

In 1975 CARE greatly increased its work in agricultural development with the provision of self-help assistance such as seeds, tools, roads, irrigation, and marketing schemes (CARE 1975). CARE's agricultural development activities were occurring in all world regions, but special efforts were beginning to take place in drought-stricken Africa. By 1976 agroforestry projects had been introduced in Niger as a way to rehabilitate badly deteriorated agricultural lands while providing wood resources (CARE 1976). And with the growing interest in the environment and its role in the development process, CARE in the early 1980s expanded its operations with the creation of a program called Renewable Natural Resources that principally focused on the integration of trees into agricultural production through agroforestry. By fiscal year 1992, CARE's ANR unit had 101 projects under way in 32 countries (CARE 1992). In the mid-1980s, CARE again expanded its operations with the creation of a new sector called Small Enterprise Management that assists small businesses. CARE's self-help development programs known as Primary Health Care, Agricultural and Natural Resources, and Small Enterprise Development rose from less than 5 percent of total development expenditures in 1961, to 25 percent in 1987, and 29 percent in 1993 (CARE annual reports; Cazier 1964, 443).[4]

The Internationalization of CARE

In 1982 CARE's organizational character was altered again when it became CARE International. The change in name reflected the growing "internationalization" of CARE's financial support. From its inception in 1945, CARE was an American enterprise with almost all of its financial support coming from donations made by the citizens of the United States or from government subsidies and contracts. This change parallels events at the governmental level, as well as the fading of U.S. domination of international development, as other industrialized nations have become increasingly active.[5]

In the late 1970's, CARE began seeking financial assistance from its former European recipients for Third World relief and development programs. In 1980, CARE Germany and Norway joined CARE Canada as two other autonomous CARE organizations (CARE 1993, 24). By fiscal year 1993, CARE International was composed of 11 members with the additions of Britain, Denmark, France, Italy, Japan, Australia, and Austria, and with a Secretariat in Brussels. In 1982 an effort was made to share the policy-making and program-planning activities of CARE, which until then had been maintained at CARE headquarters in New York. Each CARE member is allowed to establish its own implementing agency. Since CARE USA has the largest and the most experienced implementation operations, it still oversees a large portion of CARE-sponsored projects. In 1993, CARE International members supplied CARE USA with 23 percent of its contract and grant revenue income, a major source overall (CARE 1993).

CARE USA's Staff

CARE has long viewed its staff as one of its greatest assets. In his review of charities, Katz (1974, 181) remarked that "the most impressive part of the CARE offices is the nature of the people in them." Katz went on to describe the CARE staff members as determined, idealistic, compassionate, yet remarkably realistic and proud of the organization they work for. From my observations of CARE's program office and mission staffs, which date back to 1975, I would agree with Katz's general portrayal. The character of CARE's personnel, I think, can be traced to the character of the organization itself.

With CARE's successful history and character as a nonprofit, humanitarian organization, it has had few problems in attracting dedicated and excellent people to fill its ranks. Employees tend to feel good in identifying with its qualities. For example, students and recent graduates of such prestigious universities and colleges as Yale, Columbia, Harvard, and Smith donate their time to work for CARE at its headquarters or can be found in clerical and other entry-level positions at very low pay. For work overseas, CARE, in recent years, has had 5,000 qualified applicants on file for the only 150 or so positions available each year. The typical hiree for entry-level positions at CARE for overseas work has a master's degree and four years of overseas field experience (see Brechin 1989, 245). During the 1980s, which is the central period of this review, the average length of overseas experience for CARE foresters at the time of their hiring was 4.92 years, with 4.33 years in agriculture (see Brechin 1989, 244). CARE, then, tends not to hire the most technically experienced people available. Rather, it hires individuals who combine a number of qualities, such as those who are familiar with life overseas, who are willing to live relatively simply in often remote places, and who can put together programs in the field.

CARE's success in hiring dedicated and qualified people is not a recent phenomenon. The organization learned quickly from problems with employees overseas and in New York in its early years (see Cazier 1964, 60, 92, 217, 219) that a strong personnel program would have to become an integral part of CARE's effort (Cazier 1964, 93). During discussions on how to keep costs down, CARE's board of directors noted at its June 18, 1946, meeting that good "men" would work for CARE less then the premium salaries (Cazier 1964, 60). With the addition of women, this has remained true. CARE has relatively low salaries when compared with the commercial sector. This was a particular point of discussion, especially among the headquarters staff who lived in the New York City area before they moved to Atlanta in September 1993. But the nature of CARE's work keeps enough qualified and experienced staff around. Some of the other benefits of working for CARE were summed up by one long-time senior staff member: "CARE is a rough-and-tumble business. CARE is a zoo, but fun. There is great job satisfaction. Other staff members are our friends. The place is delightfully outrageous" (Brechin 1989).

CARE is a decentralized, flexible, and fairly unbureaucratic organization. Its overseas work is performed through country missions, not at headquarters. One of CARE's guiding principles, which was mentioned by several of CARE's administrators, is to hire the best people possible and then let them do their job (Brechin 1989).

CARE's Structure

Going from CARE USA to CARE International marked an important change in the organization's character and operations as well as in its structure. CARE International is a federation of independent CARE member and associate member organizations from the eleven industrialized countries mentioned above. Although each CARE member may have its own implementing unit, CARE USA's Program Department still administers a lot of CARE's overseas activities, including its work in forestry. Given this, the following description of CARE's organizational arrangements will focus on CARE-USA and its Program Department.

CARE USA

CARE USA has a board of directors, an executive staff, including CARE's president, four administrative departments, and a Washington, D.C., Liaison Office. We will review these units briefly, with the exception of the Program Department, which is discussed in some detail. The 44-member board of directors is the ultimate policy and decision maker for CARE. The board, of course, oversees the general units of CARE USA, which includes a Donor and Public Relations Department, the Washington Liaison Office, the International Service unit, and the Program Department.

CARE's Donor and Public Relations Department, as its name suggests, is charged with fund-raising activities. The general public, private corporations, and foundations are objects of this department's attention. It also promotes CARE's image and mission via media outlets through standard public relations efforts. This is not a small operation, as the Department's expenses for fiscal year 1993 were about $17 million, or 59 percent of CARE's headquarter's expenditures of $29 million. In addition, marketing provided CARE USA with a net income of $28 million for that year (CARE 1993, 40–41).[6] Slightly less than one-half of CARE headquarters' staff works in the Donor and Public Relations Department with an additional 50 or so employees located in its 13 field offices throughout the United States (Brechin 1989).

CARE's Washington Liaison Office houses its lobbyist and staff who consult with members of Congress about the distribution of government surplus commodities and other concerns essential to CARE's operation (Cazier 1964, 362). These concerns and those of government-sponsored grants remain important to CARE's existence. Finally, the International Service Unit provides a liaison function between CARE USA and the other, members of CARE International.

The Program Department

Of greatest relevance to our discussion is CARE's Program Department that oversees and coordinates the organization's country missions and their field activities throughout the developing world. The predecessor to the Program Department today dates back to 1955 when CARE established a Research and Planning Division to backstop project selections (Cazier 1964, 439). In earlier years, self-help programming was general and unsystematic. With CARE's rapid growth, especially in the area of self-help projects, CARE needed to provide some order and accountability to the projects and to assure the availability of financial resources to fund them. One event that spurred the creation of the Research and Planning Division was the sending of 10 iron lungs to a CARE mission that had only asked for one. CARE's executive director at that time, Richard W. Reuter, commented that CARE could not afford that kind of luxury (Cazier 1964, 439).

Headed by a senior vice president, the Program Department since 1984 has consisted of four administrative units named Regional Administrative Group (RAG), Technical Administrative Group (TAG), International Staff Operations Group (ISOG), and Development Education. RAG, headed by a Director of Program Administrative Assistance, provides administrative services to CARE's missions overseas. There are four regional units headed by Senior Desk Officers for Asia, East Africa, West Africa, Latin America, and the Caribbean. RAG is responsible for overseeing the budgets and backstopping the operations of country missions. One Senior Desk Officer summarized RAG's work as obtaining and juggling money, people, and projects within the specific regions. RAG administrators are responsible for overall supervision of missions and general program management, but not project management, which is left to country missions.

TAG was established in 1984 to improve CARE's technical specialization and program evaluation. It is headed by a Director of Evaluation and Sectoral Assistance. TAG has four sectoral units directed by specialists in agriculture and environment, small enterprise development, population, and health and nutrition (formally primary health care). The TAG, staff, which provides technical expertise for CARE's overseas projects, aids in the technical aspects of project planning, design, implementation, and evaluation. TAG also manages centrally funded grants and contracts.

In 1986, TAG was augmented through a $10 million USAID matching grant for rural capital formation, which established regionally located technical assistant teams along sectoral lines in Asia, Africa, and Latin America (Brechin 1989; CARE 1987a). These teams guide country missions in the design of projects, develop and hold training courses, prepare training and field manuals for CARE personnel, develop project evaluation plans, undertake internal evaluations, and assist with external evaluations. Overall, the development of Regional Technical Teams provide CARE missions with more intense and tailored local project assistance while allowing TAG in New York to focus on more general supervisory concerns.

The two other units in the Program Department are International Staff Operations Group and Development Education. ISOG is essentially the personnel arm of the Program Department and the country missions. It concentrates on staffing issues and policies and on the recruitment and training of personnel for work overseas. In contrast to ISOG, Development Education attempts to educate the general public and private donors in the United States on matters of Third World development. Unlike the Public Relations and Publication Department, which promotes only CARE and its work, Development Education attempts to be informative on the issues and problems of development in general and stresses the importance of U.S. citizens' supporting the work of organizations that promote overseas development. The unit works through publications, media presentations, and a speaker bureau.

To provide overall coordination of the Program Department activities and efficient lines of communication within the department, there is a Program Management Team composed of the Senior Vice President for Program, the directors of RAG and TAG, and the Senior Desk Officers of the regional units (Brechin 1989). In spite all of the activities at CARE headquarters, most of CARE's development programs and projects are still planned, designed, and implemented at the local level through CARE's country missions.

CARE's Decentralized Program Operations

CARE's work is highly decentralized. In fiscal year 1993 it had country missions in 53 countries in Asia, Africa, Latin America, and the Caribbean, and more recently in the former Soviet Union and Yugoslavia. CARE USA took the lead in thirty one of these countries (CARE 1993). During the period of this study, there were 230 salaried and contract employees in these countries compared with 223 in New York, for virtually a 1-to-1 ratio. When this figure for country mission and project personnel is compared to only the Program Department staff, the ratio increases significantly. For example, in forestry alone, in fiscal year 1986, 30 foresters were working on 24 CARE forestry projects. Two foresters were in New York as supervisors. Twenty-eight worked overseas. This created a field-to-headquarters ratio in forestry of 14 to 1.[7] By 1994, CARE had approximately 90 forestry projects in the field, mostly in agroforestry, with a few buffer zone management projects. For these projects, the number of CARE and host country foresters tallied over a hundred, with hundreds more involved in agroforestry extension activities (ANR 1994).

In sum, most of CARE's program staff are in the field designing new projects and implementing existing ones. Professional staff at headquarters are mostly involved in general administration and fund-raising and are not directly concerned with field activities. As is discussed in greater detail later when reviewing CARE's core technology, the Program Department for the most part only oversees and supports field mission activities and personnel. CARE's decentralized operations is an important element of CARE's character and is a defining element in its technology.

CARE's Task Environment

As noted in our review of CARE's history and character, the organization never would have survived, let alone have prospered, without the support of a number of organizations and, of course, the general public. It is the same today. Several types of organizations are most relevant to CARE's work overseas are the host governments, public and private financial contributors, and other development agencies.

Host Governments

As with the other two organizations discussed in this book, CARE's work overseas is predicated upon the formal approval of individual governments. CARE's actual working arrangement, however, is much different and is marked by greater independence. Unless CARE is simply serving as implementor for a funding agency (e.g., food distribution for USAID), it first establishes a country mission, for which the initiation may come from CARE or the host government. In setting up a new mission, CARE and the government sign what is called a Basic Agreement. The Basic Agreement formalizes CARE's operation and provides protection for mission staff. Individual projects are agreed to separately under Project Activity Agreements (PAAs).

The Basic Agreement provides the legal base from which CARE operates its overseas programs and specific country projects. Among other things, it allows for the import and export of any CARE commodities or equipment without duty or any similar tax. Although CARE employees are responsible for their personal behavior, the Basic Agreement does provide CARE and its staff with some diplomatic immunity from any legal liabilities or criminal prosecution resulting in pursuing CARE's work (Brechin 1989, 248; Cazier 1964). The Basic Agreement is generally signed by the host government's head of state or foreign minister. But more important, it assures that the equipment, commodities, and resources utilized by CARE are in fact CARE's and that they are to be distributed directly to the citizens of that country and not to the host government or its officials. Unlike the Bank or FAO Forestry, CARE USA has the power to work directly with a country's citizens and communities. People and their communities are CARE's target population, not the government or its officials. This is a crucial difference among the three organizations in their ability to reach the poorest of the poor.

PAAs are more specific and detailed contractual arrangements. They define the nature of the project to be undertaken, its target population, funding levels, and supervisory responsibilities. They tend to be negotiated and signed with host-government counterpart institutions, such as sectoral ministries. The power of approval over individual projects can expand or limit CARE's work in any specific country. By way of illustration, CARE-Belize's efforts in the mid-1980s to expand its programming in Belize by adding forestry projects was temporarily halted by the government's unwillingness to approve the project, in spite of CARE headquarters' high technical marks and approval (see Brechin 1989, 246). For CARE missions to be effective there must be a sufficient overlap between programs and projects that the government is willing to approve and support, *and* what CARE feels it can provide. This, then, raises an important difference between CARE and the World Bank and especially FAO Forestry. CARE has a say in which project it implements and even which country it works in. As a private voluntary organization, it is not obligated by international protocol to work in every nation asked of it. It has the power of choice that it can exercise to maintain control over its affairs.[8] As I emphasize in later chapters, this helps CARE maintain the quality of its projects.

Since 1967, host governments have been asked to share the cost of self-help projects (Brechin 1989, 247). Cost sharing may be in the form of local-currency contributions or (more frequently) in-kind contributions, such as the government personnel's time or use of its equipment. An important component of this cost sharing is the CARE-host government partnership in project implementation. The actual mix of CARE-host government sharing of implementation depends, in part, on the government's administrative capabilities and resources and on the type of project being implemented. To help ensure CARE's effectiveness in the field, CARE does not relinquish much (if any) control of its project to the host government's counterpart agency. This seems to be an enduring feature of CARE's approach. It explains, in part, where CARE locates its missions and what type of programs or projects are actually pursued.

Financial Contributors

CARE USA's three main sources of finances are (1) donated agricultural commodities and their shipping costs, (2) contract and grant revenue from other organizations; and (3) contributions from private donors. In fiscal year 1993, these three provided over 96 percent of CARE's revenue.

As a PVO, CARE thrives on gifts from others. Unlike FAO, however, CARE receives relatively small but highly valued donations from private citizens and foundations, as well as from governments. Because CARE has direct ties to its voluntary donors in a very competitive environment, its officials worry about its reputation as an efficient and effective development organization. In the little literature that exists on CARE, it was emphasized repeatedly that CARE officials have always remained steadfast in their belief that the public will not make financial contributions to an inefficient and ineffective organization (Campbell 1990; Cazier 1964; Katz 1974). Still, these concerns certainly must be more ecumenical than those that CARE may have about its general reputation or public image. It would be difficult for the average donor to know how well any particular project or even program was faring. Images are key, but usually have some factual reality.

CARE has for years carefully cultivated a long list of loyal supporters from the public at large (Campbell 1990; see Cazier 1964; Katz 1974). Although its overall total contribution is relatively small (10% in fiscal year 1993), the general public, of course, has always remained an essential source of financial support. Each year over 1.5 million individuals from around the world make financial contributions to CARE. Donations range from one dollar to several thousand and help maintain the independent nature of CARE's operation (Brechin 1989; CARE 1986a, 10). This cannot be overemphasized. Without the support of private groups and the millions of private citizens, CARE would need to depend even more upon contracts with other organizations. Individual donations also help to maintain both the myth and reality of CARE's independent and private status.

Donations from the general public and from private corporations provide CARE with its primary source of *unrestricted* funds. Thus, CARE has resources of its own that allow it to remain flexible and innovative. With those resources, CARE can be more independently responsive to new program initiatives or can continue existing programs until new funding can be arranged. It is also an important financial source for overhead costs. And private donations support most of CARE's self-help programs, including forestry.

Most of its financial resources come from other sources, in particular the U.S. government, as well as from other national and international funding agencies. The U.S. government is without question the largest single source of revenue for CARE USA. In fiscal year 1993, 50 percent of CARE's total revenue came from donated agricultural commodities and their shipping costs, the bulk provided by the U.S. government. Contracts and grant revenue were about 35 percent of CARE's revenue for that same year. Here again, funding from the U.S. government (largely through USAID) accounted for about 42 percent of that total. Other sources of contracts were from CARE International members (23%), host governments (15%), the United Nations (9%) and other governments (6%) (CARE 1993).

CARE has a long history of working with international organizations. CARE developed a book program with UNESCO in the 1950s (see Cazier 1964). The World Bank has supported a major CARE school construction project in Liberia, and, for a number of years until the late 1980s, provided partial funding (along with the African Development Bank) for a CARE feeder roads program in Sierra Leone (CARE 1986a; CARE 1987a). In the late 1980s, the Bank and CARE collaborated on a community forestry project in Uganda as a subcomponent of a much larger forestry project (CARE 1987a, 16). The European Economic Community has supported CARE's African emergency efforts with contributions of food and funding for Mauritania and the Chadian Agricultural Cooperative (CARE 1986a, 11).

During most of the 1980s, CARE received its largest support among international organizations from the United Nations High Commissioner for Refugees (UNHCR). From fiscal year 1979 through fiscal year 1985, UNHCR provided CARE with over $44 million for relief projects for refugees in Somalia, Thailand, and Sudan (CARE 1986a, 11). Other international organizations supporting CARE's activities have been the United Nations International Children's Emergency Fund (UNICEF) and the World Food Programme. These organizations have supported wide-ranging CARE efforts in Bangladesh, India, the Philippines, Chad, and Haiti (CARE 1986a).

Other Development Organizations

CARE operates largely in a world of other development and relief organizations. It is not, as one might expect, always a highly competitive world with organizations attempting to outcompete one another. Each organization tends to have its own niche in programs, countries, and funding sources, although the area of general voluntary giving can be quite competitive. There is considerable effort, however, not to overlap the activities of other organizations because it is viewed as wasteful and disrespectful of another's turf.

Instead of conflict out in the field, cooperation is often the norm, as project arrangements are frequently worked out among the different organizations. For example, a number of the PVOs, such as the Catholic Relief Services, Lutheran World Relief, and United Jewish Appeal, are religious and get a good chunk of their funding through religious affiliations. This is not to suggest that these organizations do not overlap countries and programs, but there is so much work to be done that their activities tend to be more complementary than competitive. By way of illustration, Beryl Levinger, a former CARE Assistant Executive Director for Programs, noted in the 1985 *Annual Report* that in providing relief to the Ethiopians in Eastern Ethiopia (Dire Dawa) there was significant collaboration among the relief organizations working there. CARE used weight/height charts provided by Save the Children/United Kingdom. Lutheran World Relief Federation provided CARE with support in incorporating a primary health care component into CARE's feeding activities, and Catholic Relief Services shared its stock of bulgur wheat to help CARE start its feeding program. CARE, in turn, provided the others with some its of corn/soy/milk stock as an element in supplementary feeding programs.

One development organization that has a great deal of interaction with CARE and its projects is the U.S. Peace Corps (Campbell 1990). Since its creation in 1961, Peace Corps volunteers have worked on CARE projects overseas. In fact, CARE was asked to train the original volunteers (Campbell 1990; Cazier 1964; Kaufman 1971). In addition to providing work for the volunteers, CARE receives an important source of free labor for its projects. This free labor helps CARE to maintain high staff intensity at its project sites. More important, perhaps, the Peace Corps provides CARE, and the other development organizations, an important pool of personnel with overseas experiences from which to fill its own ranks. Many of CARE's Program Department professionals are former Peace Corp volunteers, along with about one-third of its overseas staff. With the growing numbers of Peace Corp volunteers, about 3,000 annually during the 1980s, CARE has had a large pool of talented people from which to select (Brechin 1989, 254).

CARE's task environment is largely supportive and unconstraining. Its private humanitarian nature and its reputation for efficiency and effectiveness seem to have provided it with goodwill and financial support from citizens, foundations, corporate donors, and U.S. and foreign governments. Combined, these attributes have given CARE its considerable independence and flexibility. Unlike the more political environment of FAO, CARE can and does refuse work asked of it, and, unlike the Bank, it is not constrained by having to package its work in loan agreements.

CARE's Core Technology

CARE's purpose, as stated in its 1993 *Annual Report,* is: "to help the developing world's poor in their efforts to achieve social and economic well-being. We support processes that create competence and become self-sustaining over time. Our task is to reach new standards of excellence in offering technical assistance, training, food and other material resources, and management in combinations appropriate to local needs and priorities. We also advocate public policies and programs that support these ends."

CARE's approach to its development work can best be summarized as flexible, decentralized decision making (see Tendler 1982). Here we have CARE's structure influencing its core technology. CARE combines its mandate of working with people in poor communities in poor countries with decentralized professional expertise on the ground to promote community and rural development activities. Outside of emergency aid and related food programs, CARE places special emphasis on the principle of self-help at the local community level. It is an organization that tries to work directly with impoverished people to better organize and use their own human and material resources toward socioeconomic improvement in ways that are meaningful to them (Brechin 1989). CARE's programs conform to the development plans of host governments. CARE utilizes a "partnership approach" in projects with appropriate host government counterparts. As discussed above, CARE's projects are often financed by bilateral or multilateral organizations, but it has its own resources to contribute as needed. In keeping with its humanitarian mission, CARE's projects are grants to local people which do not have to be repaid. The partnership arrangement, however, does require the host government to make at least nominal contributions to the projects either in cash or, more commonly, in kind (e.g., staff time and equipment use).

Programming Operations

To guide its decentralized decision-making technology, CARE has established particular programming criteria and operations procedures.[9] CARE's programming criteria promote six basic objectives: scope, fundamental change, participation, feasibility, assimilability, replicability, and clustering (CARE 1986a). With its work focused on helping the poor of the developing world, CARE believes its programming, to be effective, should be of sufficient *scope,* affecting meaningful numbers of the poor. Another programming objective is to improve peoples' lives in appropriate ways that promote *fundamental change* for people's direct benefit. CARE's self-help approach is expected to stimulate greater individual and community self-sufficiency. Essential to the process is *participation.* The people affected by the project must be a part of the project for it to be truly meaningful. In addition, local participation not only contributes to project cost effectiveness but, more important, it generates commitment to the project, confirms the fact that the project belongs to the people, and provides communities with new skills in self-reliance. A fourth programming point is *feasibility.* To encourage sustainable development, the project should be an appropriate response to real needs. And it must be a project that can be sustained eventually by local people with their own resources. *Assimilability* is the fifth point. CARE's activities should be appropriate and designed realistically in order to allow host government counterparts to institutionalize—and thus continue—the program after CARE's support has ended. CARE's project concepts also should be *replicable.* CARE officials feel that development can best be assisted when good ideas can be expanded and promoted elsewhere in the region or country. Replicability is an important way to maximize the investment associated with any given program or project. Finally, *clustering* suggests that several complementary development projects should be placed together to intensify their collective impacts on the target communities, maximizing support services while minimizing administrative costs.

During the 1980s, CARE organized its work overseas through its established network of decentralized and semiautonomous country missions that operate in concert with CARE's Program Office in New York. Together the missions operate through a multiyear planning system that has some centralized control, allowing CARE to plan projects years in advance with host government counterparts and funding agencies, yet remaining flexible and responsive to local conditions and new initiatives.

An important characteristic of CARE was its work through, for operational and planning purposes, permanent missions, functioning under the arrangements of the Basic Agreement already noted. The missions administer development *programs* that usually consist of a number of projects, often across several sectors. This is in contrast to the World Bank and FAO, which are typically focused temporarily on single projects.

Since about 1973 and through most of the era discussed in this book, CARE operated its overseas program by using a management system called Multi-Year Planning (MYP) (Brechin 1989; Campbell 1990; CARE 1979a). This system was an integrated programming approach that provided built-in mechanisms for planning, implementation, and evaluation. It was created to better coordinate CARE's field activities with headquarters, which oversees human and financial resources without stifling the decentralized activities of the missions. Overall, the MYP tried to established a more uniform and routine system for administering resources. It has allowed for more effective planning by presenting CARE's programming activities and needs on a three-year basis. Every year each mission director submits MYP plans to headquarters for review and approval. Submitting the plans yearly allows for continuous monitoring and adjustments to ongoing projects as well as the development of new projects. More recently, the MYP program is being replaced by the Long Range Strategic Plan (LRSP) to make the field-headquarters' link even more efficient.

Salient Features

One of the most salient features of CARE's work has been its use of country missions. Through time, it has been able to acquire a wealth of firsthand knowledge useful for its work. From these vantage points, CARE staff members have been in a strategic position to better understand the country where they work, its people and their needs, the government and how it works, and what types of projects would likely work best and where. Of great importance has been that CARE itself frequently becomes a known quantity, often trusted and respected by government officials and local people alike. With administrative structures and staff already in place, the CARE staff has often been able to concentrate on programmatic efforts and be in a position to develop or quickly accept new project or program ideas.

CARE's missions have been viewed as semiautonomous units. "Field missions are very autonomous and decentralized, and they *should be,*" responded one CARE Program Department administrator I interviewed. "Missions get their way 98 percent of the time" (Brechin 1989). The staff members of RAG, who are directly responsible for working with the missions, tend to view their job as assisting the missions, not controlling them. Headquarters' role is to provide a more objective look at project proposals and at the missions' financial and human resources. This is not to suggest that headquarters does not have superior authority. It does. It approves each new project, allocates financial and other resources to the missions, and evaluates the performance of mission staff. Even so, it still leaves important implementing decisions to the CARE missions. Once headquarters accepts a project proposal on its technical merits and assigns funds, the proposal is really little more than a "hunting license" to proceed with the project. A "hunting license," as opposed to a project "blueprint" found at the Bank, gives CARE missions considerable flexibility in the administration and implementation of its projects.

CARE's projects are not etched in stone. This is a key part of its decentralized decision-making core technology. With a "hunting license," CARE field personnel are only required to put something together that resembles the project proposal; the basic instructions are, "Go make it work" (Brechin 1989, 258). This flexibility gives CARE field personnel opportunities to better tailor project features to meet unforeseen or changing conditions and to hire their own project staff. In addition, most of CARE's project ideas have been generated at the mission level, which helps to assure that the projects are feasible, appropriate, and well adopted to the country context.

CARE's "bottom-up" programming and decision making at the mission level, along with the availability of internal sources of funding, go a long way in explaining CARE's general success at implementing its development projects. In addition, for CARE external funding often comes from successful programs and projects, not the other way around. Or stated another way: Dollars generally don't drive programming; good programming is often followed with more dollars. Success breeds success since good ideas that are well executed are more likely to be supported by funding agencies. With CARE's presence overseas, and its interest in developing new program and project ideas, CARE is better poised to detect unmet needs that emerge from local people, communities, or host governments. And with its own (albeit limited) sources of revenue, CARE can be an innovator and experimenter by testing new ideas on the ground without depending on external funding. This is precisely how CARE's work in community forestry began. CARE's early self-funded experience in the Majjia Valley eventually led to the expansion of CARE's forestry program and multimillion-dollar support from external sources—in particular, USAID.

CARE missions certainly have project failures too. Its projects, however, are almost always small in scale, especially new ones; so if they do not succeed, any negative consequences are probably negligible—which would not be the case in a World Bank project. CARE's decentralized decision-making processes and the associated monitoring technologies of its field missions allow it to detect failures more quickly and learn from them. With the permanent nature of the missions, and number of projects going on, it is relatively easy for a CARE mission to discontinue unsuccessful projects without seriously affecting CARE's overall operations in that country. This is much less true for an organization whose presence in a particular country is based solely upon a particular project, as is often the case with the Bank and FAO. This ability is also enhanced by CARE's control over project resources. Changes do not require the degree of formal negotiation with other parties that is required by the Bank and FAO Forestry.

Another salient feature of CARE's work is its focus on people and the community. This focus is a consequence of CARE's history as a relief organization. Even through its expansion, CARE has continually attempted to do good works directly for less fortunate people in the developing world, *not* for their governments. CARE's people-to-people community development approach has influenced its development programming and projects in a number of ways. For example, CARE officials believe that a rigid benefit-cost analysis is inappropriate for the humanitarian nature of its work. The officials are more interested in helping people provide for their basic needs than in determining the appropriateness of the service based on an economic formula. CARE projects are subjected to a review process based on the need to be addressed, field experience (and even intuition), availability of CARE resources to do the job, and the technical soundness of the proposed project. Economic formulas do have a role, but they are of a different kind. An important consideration in project selection is financial leverage, that is, how much additional revenue for the project can be generated for every dollar provided by CARE. For example, unless it were a new program idea being tested, a project that can generate a higher ratio of external dollars to CARE dollars would likely be selected, among equally appropriate alternatives, instead of a project that generated only a 1-to-1 ratio. Obviously, high leverage ratios stretch CARE's resources and allow CARE to do more work in more places. Another consideration is the actual cost of working in a particular country. Projects in countries that are expensive to work in are more likely to be scrutinized by headquarters. Decisions will be made in part on whether CARE's resources could be more effectively used elsewhere. Through such a decision process, CARE tries to maximize the effects of its work.

Another important feature of CARE is the nonpolitical nature of its work. CARE is organized around the elimination of human misery. Its efforts in this regard, however, are not without limits. As a humanitarian organization, it does not dwell on highly political issues such as human rights. Its niche has been marked by technical and material assistance and emergency relief to the poor people of the poorest countries. This is not to say that CARE's efforts could not be construed as political at times. Cazier (1964) gave detailed accounts of how some of CARE's relief work—thinly veiled "clandestine" action in the war against communism—brought favors to the America people and the U.S. government. But its work with the United States Information Agency in distributing pro-American reading materials was highly criticized by the public as being too closely aligned with the interests of the U.S. government. As a consequence, it was forced to abandon the program (Brechin 1989; see Cazier 1964). Although the U.S. government has always played an important part in supporting CARE's efforts, CARE has, on the whole, been able to maintain an independent identity as private humanitarian organization. Its separate identity, however, is intricately linked to CARE's own self-interest. CARE's fund-raising efforts depend largely on its reputation. Private donors have tended to respond well to CARE's independent character and humanitarian appeal.

In addition, CARE only goes where it is formally invited—but then, it does not accept every invitation. CARE has long been considered an American organization, but with its recent internationalization, it will become much less so in the future. For example, CARE has not only worked with U.S. friends such as Honduras, Guatemala, and Costa Rica, but it has also worked with Nicaragua (during the 1980s) when it was not a friend of the United States. It has offices in Mozambique and Angola as well. Its mission in Afghanistan was temporarily suspended with the arrival of Soviet troops in 1980, but so were its missions in El Salvador and Uganda without Soviet troops. Politics has certainly affected CARE's activities, but it has not dominated them. In this sense, it is not a political organization. CARE's nonpolitical, humanitarian character has positioned it to work in a number of countries much more quietly and effectively than other organizations have.

Assessment

CARE's decentralized decision-making technology has helped to make the organization efficient, effective, and flexible. And as mentioned earlier, its private, humanitarian character has opened many doors for CARE and given it considerable goodwill. Still, its technology and character have also severely limited the nature and scope of its activities. For example, it can only work with the poorer segments of the population in the world's poorest countries. Although not in itself necessarily bad, it does limit its operations. As a result, it is problematic for CARE to help the poor of Europe (unless under emergency situations as found in Bosnia-Herzegovina) or the working class in Latin America or in Asia's wealthier countries.[10] In 1993, CARE missions were located in only 46 of the world's roughly 84 poorer developing countries (see CARE 1993; World Bank, 1992a). CARE is an international organization with quite a limited geographic focus, in contrast to the Bank and FAO, which work in many more countries.

Perhaps more crucial is that CARE works only at the local level, most often with generally powerless people on small-scale application projects, frequently in isolated areas. Although it generally does good work, critics claim that small isolated projects more often than not create temporary, isolated pockets of development that are eventually washed away or diminished by the more-powerful and persistent uncomplimentary waves of national-level policies (or lack of them) (Annis 1987; Brechin 1989; World Bank 1991a, 1991b). Sustainable development, it is now often argued, can only be attained by the proper policies and action from the national government (see Feinberg and Helleiner 1986; World Bank 1991a, 1991b). It is not that "bottom-up" development projects are unimportant, but rather that to be effective they must be complemented by "top-down" activities (Annis 1987; Ayres 1983; Brechin and West 1990; Leach and Mearns 1988). CARE lacks the resources and the working arrangements of a World Bank, or the political legitimacy of an FAO, to influence the policies and administrative actions of national governments.[11]

Summary of CARE's Character and Technology

From its beginnings in 1945, CARE has evolved from a small, designated-package voluntary relief agency servicing only Europe to a relatively large, highly flexible, efficient, and sophisticated self-help development and relief organization operating in over 50 countries (CARE 1993; Cazier 1964, 444). Throughout Campbell's (1990) and Cazier's (1964) extensive reviews of CARE's history, both authors repeatedly stress its efficiency, effectiveness, flexibility, innovativeness, and popularity.

One of the major reasons for CARE's popularity at home and abroad has been its humanitarian and nonreligious nature. With the issue of the separation of Church and State, and the desire to maintain broad-based political support, the U.S. government has always looked favorably upon CARE. Very early in its existence, CARE was considered by many as the relief and development organization of the American people, not of its government (Cazier 1964, 84). Likewise, a number of recipient governments disapprove of the proselytizing that was often attached to the relief administered by religious organizations. CARE has avoided that problem with its broad, nonsectarian support. It also has worked to dissipate any political overtones to its operations, in spite of its substantial governmental support, by focusing only on the technical and humanitarian issues of events in a people-to-people manner (Cazier 1964, 315, 324). With its internationalization in 1982, CARE has enhanced its status as an independent organization, one no longer simply American.

Because of its voluntary nature, CARE has always felt it had to especially concern itself with its efficiency and effectiveness. Simply put, CARE has believed, and no doubt correctly, that the public will not repeatedly give money to an organization viewed as inefficient and ineffective. In their histories of CARE, Campbell and especially Cazier frequently mention the concern that both the board and the public have expressed on this issue. Likewise, the U.S. government might likely find it politically difficult to justify awarding contracts to a PVO with a poor reputation. As a consequence, CARE has always taken active measures to present itself as a sound organization. Although CARE falls short of its near flawless image, it has taken serious measures to control its administrative costs at headquarters and has promoted effectiveness overseas. To this day, CARE prides itself on having one of the lowest overhead expenditures in the charity business (see Brechin 1989, 265; Campbell 1990; Cazier 1964), although this by itself does not necessarily make it a better organization. Such a fact is of interest to donors who want to see their contributions feeding the poor, not a bureaucracy.

From its very beginning, CARE has worried about its efficiency. To assure a skeptical public, CARE provided copies of signed receipts to prove that packages were received by the people to whom they were sent. CARE has always preferred to control all aspects of its work overseas. As noted by Cazier (1964), "Very few agencies, if any, could duplicate CARE's pattern of supervision of overseas operations by U.S. personnel" (323). Experts still remark on the extraordinary staff "intensity" that CARE provides for its projects overseas (Brechin 1989, 265). To capture the attention and imagination of the American public and to expand its work overseas, CARE had to provide new and appropriate services wherever there was sufficient need. It has also relied on its public relations machine to put a positive spin on its work to keep private donations coming in. Still, even good public relations cannot turn a sow's ear into a silk purse. CARE's decentralized decision-making technology and tight control over its activities has been indispensable in allowing it to keep its best face forward. On the whole, CARE's performance has been flexible, responsive, and innovative.

Chapter 4

Patterns in the Forestry Programs: A Descriptive Account

In this chapter, we review the forestry activities of each organization. In particular, we see how well each stacks up against three of the four review criteria presented in the Introduction: commitment to rural development forestry, appropriateness of the work, and general effectiveness. The fourth criterion, program entrance, was already reviewed in the Introduction and is discussed more theoretically in the final chapter.

This will be mostly a descriptive discussion and is thus likely to be of interest to those intrigued by the forestry activities of the World Bank, FAO, and CARE. The findings presented here do provide the empirical foundation for the more conceptual and theoretical exchanges that follow in the later chapters.

The World Bank's Forestry Program, 1969–1992

I want to describe the particular patterns that reflect the World Bank's lending for forestry activities from 1969 to 1992. I pay particular attention to the Bank's work in rural development forestry where the data illustrate not only the overall size of the Bank's investment in this kind of forestry but also the regional and country distribution of this investment. Several comparisons are made to the Bank's lending for industrial forestry activities. We begin by reviewing the Bank's general lending for forestry activities across time.

During the period in question, the Bank made loan agreements totaling nearly $2.8 billion (or $3.8 billion adjusted to 1992 U.S. dollars) on forestry activities (Brechin 1994). About $1.1 billion, or slightly more than 41 percent of the total, was allocated to rural development forestry. Another $1.2 billion, or 43 percent, was loaned to countries for traditional forestry activities. The remainder went to mixed projects containing both traditional and rural development forestry. The yearly loan totals for these forestry projects are presented in figure 4.1 (a and b).

A review of figure 4.1 (a and b) shows the total funding in all types of forestry, using both normal and adjusted 1992 dollar values. The value of all loans and credits reached their highest level in fiscal year 1990, totaling $536 million. The previous high had been slightly more than $325 million in fiscal year 1985. In 1992, the second highest year ever, more than $458 million was given out in loans and credits.[1] The figures do demonstrate an increase in the Bank's lending for forestry, resulting largely because of greater attention being paid to rural development (in the late 1970s) and mixed forestry projects (in the 1980s).

A breakdown by world region of Bank lending for industrial and rural development forestry (fig. 4.2) shows that nearly 49 percent of the Bank's forestry portfolio went to Asian countries. Africa was second with 28 percent. Latin America and the Caribbean (LAC) and Europe, the Middle East, and North Africa (EMENA) had almost 12 percent of the portfolio each. One of the more interesting findings is that, although Africa was second in total size of its forestry loan portfolio, it led all regions in number of loans (42) and number of countries receiving loans (19). Asia was second with 36 forestry loans to 10 countries. This suggests that African countries tended to receive smaller loans on average than Asian countries, a point to which I will return.

When we look only at rural development forestry, with the first loan in 1974 (see fig. 4.3), more than half (51.2%), some $583 million, went to Asia. Latin America and the Caribbean followed, Africa was third, and EMENA consumed the remainder. A more striking finding is that, within Asia, 80 percent of the rural development forestry loans and credits went to India. Brazil dominated the LAC region with about 69 percent of that region's total. Africa's smaller portion, however, was more evenly distributed among 10 countries.

India not only dominated the Bank's work in Asia, but it received 56.9 percent of all the Bank's rural development forestry loans from 1978 to 1986. From 1987 to 1992, a lower but still significant amount, 23.4 percent, of the Bank's total effort went to India. The second leading country was Brazil with $167 million. Brazil was a recent addition to this type of forestry and

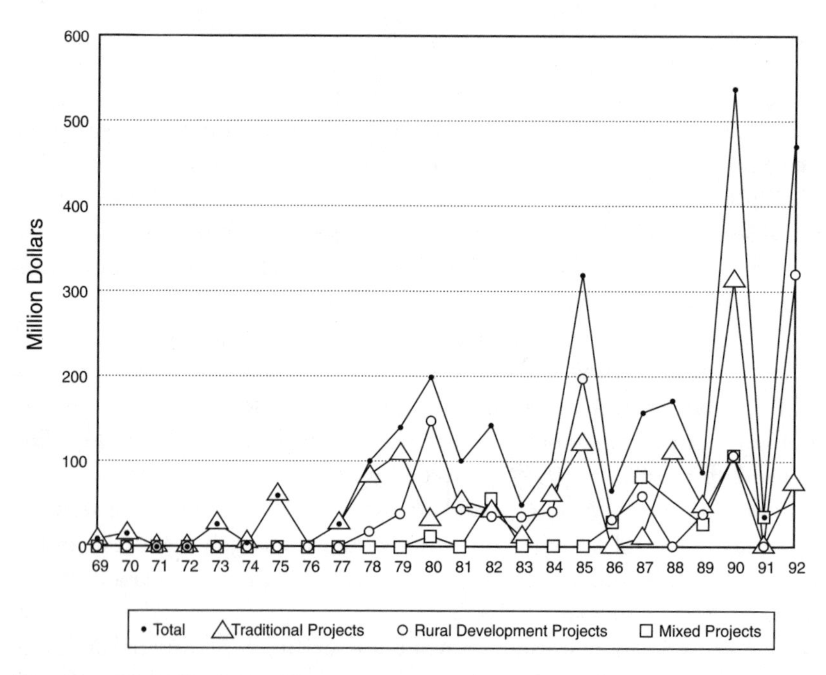

Fig. 4.1a. World Bank Total Loans in Forestry by Type, 1969–1992 (Nominal Values)

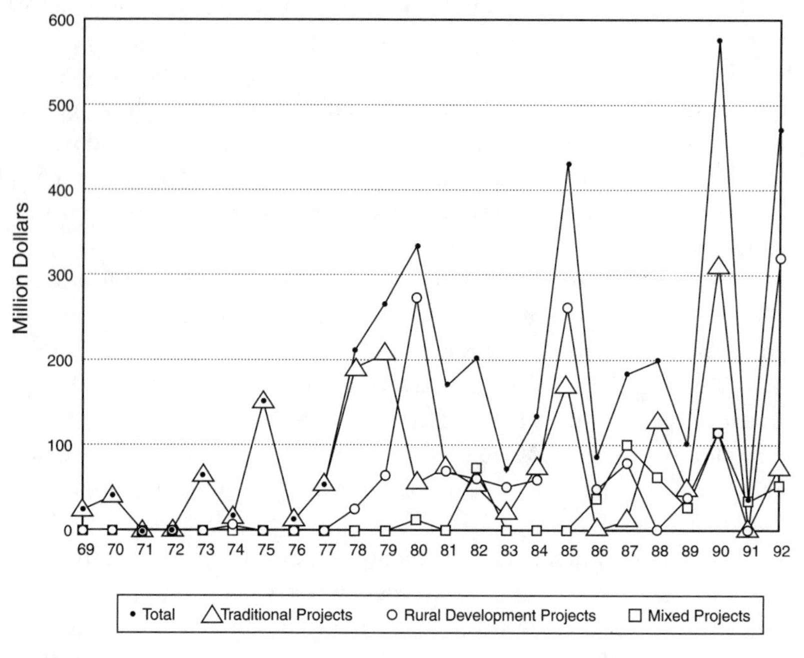

Fig. 4.1b. World Bank Total Loans in Forestry by Type, 1969–1992 (1992 Dollars)

had one loan approved in 1992. Within Africa, the division among the 10 countries was more equal. Ethiopia led the continent with $45 million in rural development forestry loans or, 21 percent of the African total. Rwanda was second with $35.1 million. From these data we see that India alone received more loans for rural development forestry than all other Bank countries combined, and twice the regional totals of either Latin America and the Caribbean or Africa. I provide explanations for this curious pattern later in this chapter.

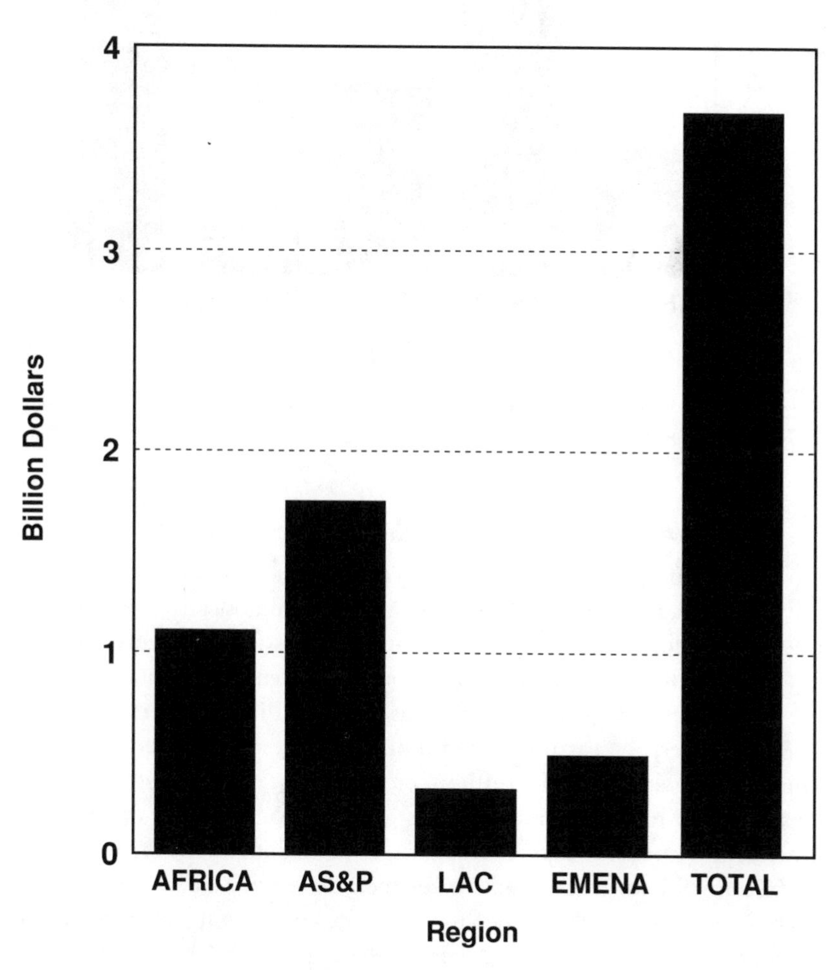

Fig. 4.2. World Bank Total Forestry Loans by Region, 1969–1992 (1992 Dollars)

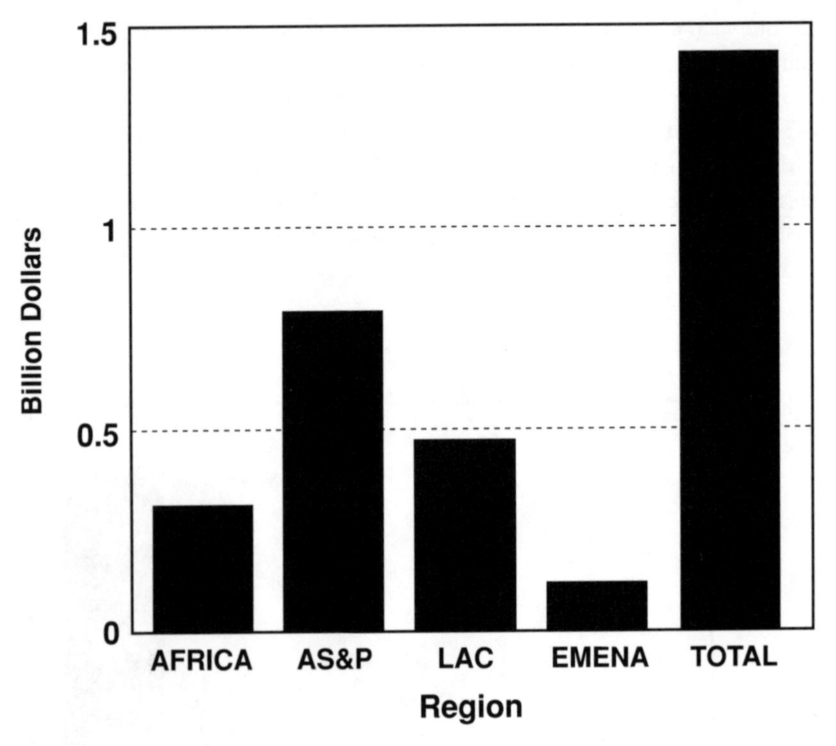

Fig. 4.3. Total World Bank Lending in Rural Development Forestry by Region, 1974–1992 (1992 Dollars)

The Bank's Commitment to Rural Development Forestry

When comparing traditional with rural development forestry projects in the early years, from fiscal year 1969 to 1977, almost all were geared to traditional efforts. There were 10 traditional forestry project loans valued at $121.7 million, which represents about 98 percent of the total. The single rural development forestry loan was for $2 million. Between 1978 and 1986, however, the nature of the funding for forestry changed markedly. During that period, $550.4 million, or 44 percent of all forestry projects, were invested for traditional purposes, while $605.9 million, or 48.5 percent, went to rural development forestry. The remainder went to mixed projects. From 1987 to 1992, overall investment in forestry continued to climb with $561 million (39%) loaned for traditional projects and $529.8 million (37%) for rural development. Again, the remainder was devoted to mixed projects. Here we see a sustained effort to support rural development forestry, with a dramatic increase in mixed projects. Keep in mind, however, that the Bank's work in forestry is minuscule when compared with its total development loans for all sectors. In 1992, the figure for development loans was $21.7 billion; of this, only 2.1 percent went to forestry activities of all types combined (see Brechin 1994; World Bank 1992a).[2]

A review of the numbers shows that the percentage of rural development forestry compared with total Bank financing of all projects varies from a low of 0.05 percent in 1974 to 1.37 percent in 1985 and to a high of 1.53 percent in 1992 (Brechin 1994; World Bank annual reports).

Appropriateness of the Bank's Rural Development Forestry Program

Here we see how much of the Bank's work in rural development forestry has gone to those countries where it was most desperately needed (as identified by FAO). Table 4.1 lists 25 countries with acute scarcity in fuelwood supplies, and the Bank's rural development forestry work in those countries.[3] From 1978 to 1992, the Bank financed 17 rural development forestry projects in 9 of the 25 countries listed. Overall, $290 million, or only 26.4 percent, of the Bank's rural development forestry loans and credits, and 10.4 percent of the Bank's total loans in forestry, went to those countries in acute need.

Table 4.1 also keys the Bank's forestry work in those countries that have a *critical scarcity of fuelwood and that obtain over two-thirds of their energy supplies from wood.* The Bank financed 15 rural development projects in 8 of the 12 neediest countries. Only $273 million in loans and credits, or about 25 percent of the Bank's rural development forestry program, and less than 10 percent of its total forestry program, were directed to these countries.

Effectiveness of the Forestry Program

In 1978, the Bank published its first forestry sector policy paper (World Bank 1978). The Paper announced the Bank's newfound interest in forestry investments. In particular, it identified the need to expand the concept of Third World forestry investment beyond the scope of traditional industrial forestry to include a range of neglected forestry-related activities of the rural poor. The production of fuelwood and fodder, sound agricultural management, and local forest industries for rural areas were to be the central components of the Bank's forestry lending. The Bank planned to allocate 60 percent of its forestry-related investments to these nontraditional purposes (World Bank 1978, 52).

Table 4.1
Appropriateness of the World Bank's Work in Rural Development Forestry: Bank Forestry Projects Compared to Countries with Acute Scarcity in Fuelwood
Fiscal Years 1978–1992

Countries	*$ Amount (millions)**		*No. of Projects†*	*% Total Rural Development Forestry*	*% Total Forestry*
	Nominal	*Adjusted*			
Botswana					
Burkina Fasoz‡	14.5	24.7	1		
Burundi	17.1	24.7	2		
Cape Verde					
Chad‡					
Comoros					
Djibouti					
Ethiopia‡	45.0	55.6	1		
Kenya‡	57.4	75.0	2		
Lesotho					
Mali‡	10.8	16.7	2		
Mauritania					
Mauritius					
Niger‡	14.6	24.4	2		
Rwanda‡	35.1	53.2	2		
Somalia‡					
Sudan‡					
Swaziland					
Afghanistan‡					
Nepal‡	65.5	87.8	3		
Bolivia					
El Salvador					
Haiti‡	30.1	31.9	2		
Jamaica					
Peru					
	290.1	394.0	17	26.4	10.4§
	273.0	369.3	15	24.8	9.75‖

Source: Data compiled from Brechin, "Forestry Project Data for the World Bank, FAO, and CARE International, 1969–1992" (1994, mimeograph).
*Dollar figures are nominal and adjusted to 1992 dollars.
†In an effort to be conservative, mixed projects have been included in the totals.
‡Indicates a country that depended on fuelwood for over two-thirds of its total primary energy consumption in early 1980s (see *Tropical Forestry Action Plan,* (1985) 49).
§Data in this row reflect the Bank's forestry projects in 9 of the 25 countries listed in the table.
‖Data in this row reflect the Bank's forestry projects in 8 of the 12 countries listed in this table with the most acute fuelwood needs (i.e., those countries that used fuelwood for at least two-thirds of their energy needs).

Although various analyses (including the Bank's own) showed that it did not meet these targets (Brechin 1994; World Bank 1991a, 1991b), the publication of the original sector paper itself generated important international attention to this long-neglected area. It helped to provide the legitimacy and financial support needed to make the concept of community and rural development forestry a reality. This clearly demonstrates the Bank's ability to establish its own policy initiatives and, in the process, to change the direction of international forestry. Despite the failure to meet its goals, the Bank's 1991 forestry policy sector paper continued its commitment to rural development forestry.

One of the most interesting aspects of the Bank's efforts, however, is that it illustrates well the adage: "The spirit is willing, but the flesh is weak." Although many believe the Bank's intentions were indeed honorable, it has proved almost completely incapable of putting into operation its own policy of supporting the rural poor (Brechin 1989; World Bank 1991a, 1991b). Especially during the late 1970s and throughout the 1980s, many of the Bank's forestry projects labeled as rural development essentially became new forms of industrial forestry controlled by governments, which benefited mostly urban citizens, the rural rich, or the government. The rural poor, the intended beneficiaries, were largely bypassed. The reasons that shaped the Bank's performance are complex, but I argue that it was largely the Bank's investment-related core technology that was fundamentally at fault, especially when its work was *unconstrained* by other institutional arrangements. The Bank appears to perform most admirably when limited simply to the role of project financier (Shiva 1991).

The Bank's poorest performance clearly occurred in Africa, a region where, ironically, its efforts were most desperately needed. Here the Bank's influence over recipient nations was high, the continent's institutional capacity weak, and political commitment to community and rural development forestry minimal. This created an untenable situation, and the goals of community forestry quickly became convoluted. In these situations, the Bank controlled the project designs and seemed unable or unwilling to incorporate all the essential social components into them. As a result, the projects tended to be seriously flawed, and performance (and the rural poor) suffered accordingly.

Until the late 1980s, the Bank's rural development forestry projects in Africa typically were government-owned fuelwood plantations. These large-scale plantations, often located outside the country's capital city, were created to produce fuelwood and poles for urban dwellers, not rural families. Many of the projects had a nursery component that was geared to distributing seedlings free of charge or at nominal costs to local farmers. Many of the seedlings, however, were eucalyptus, an exotic, which has the advantage of being a fast-growing, relatively maintenance-free species that can grow in semiarid and other marginal lands. But eucalyptus is not a preferred firewood species, almost useless for animal fodder, and can damage agricultural production and the local environment (Shiva 1991).

As a result, the projects were plagued with problems. The fuelwood plantations in Niger, for example, were often referred to by international foresters as the most disastrous forestry project of the Sahel in the 1980s. Here, the plantations were irrigated from the Niger River. Production costs were much higher than expected and the yields lower (see Brechin 1989; Heermans 1987). The project directed fuelwood and poles to the urban dwellers of Niamey, the capital, not to rural areas, but to little avail. The prices of the wood grown in Niger were 11 times the going rate for fuelwood and 4 times the rate for construction poles. Similar projects with similar results were financed in Mali, Burkina Faso, Senegal, Malawi, and Rwanda.

The similarities were not a matter of chance. Once a new type of project makes it through the difficult project pipeline at the Bank, it serves as a template of a "bankable" project. It is then copied by the other Bank staff out of simple logic: if a given project format made it through the bureaucracy successfully, then it is probable that another one will as well. It is less risky and more efficient for staff to follow the template than to experiment with a new package. Thus, the same flaws were replicated. Additionally, most of the subsequent projects were begun before the results of the original one were known.[4] Little emphasis or concern was given to the actual performance of these projects until they were funded, not only once but frequently twice—that is, not until the money had been moved. The Bank's willingness to alter its designs came only after the projects' poor results became public knowledge.

The Burkina Faso and Malawi projects provide other illustrations of the inadequacies of the Bank's efforts. In Burkina Faso, a relatively large and rare natural forest was bulldozed to make way for the construction of a government-owned, completely fenced, fuelwood plantation. Villagers of the adjacent communities had used the forest as important sources of fuelwood and other forest-based products. The forest was destroyed under the accepted assumption of the time, namely, that plantations had greater economic efficiencies than natural forests. Of course, the natural forests provided numerous benefits to local villagers, many of which were not part of the Bank's calculations. With traumatic results and after severe criticisms, the Bank has since changed its assessments of both plantation and natural forests (World Bank 1991a, 1991b).

In Malawi, a project that during the early 1980s attempted to produce seedlings for small farmers, in addition to a fuelwood plantation, was totally ineffective. The decentralized system of 88 nurseries produced nearly 900,000 seedlings each year (French 1985, 105). It also had an extension subcomponent to assist rural families in the establishment of family woodlots. Again, the production costs of the fuelwood for the urban areas were 10 times greater than the existing price of firewood. The demand for seedlings was grossly overestimated since the existing indigenous seedling market was being satisfied from natural stocks. The highly subsidized seedlings remained unused in the nurseries. And finally, the extension service was geared toward pushing the unwanted seedlings, not to disseminating the kind of tree-growing information desired by the farmers (French 1985, 105, 112).

Assessing the performance of the Bank in India, however, is not as straightforward a case as in Africa. As indicated earlier, India has dominated the Bank's lending in community forestry, receiving about 41 percent of the Bank's total rural development forestry loans from 1978 to 1992. It was also a prominent customer, which allowed it to manipulate the Bank. When discussing the Bank's work in India, it is important to remember that, unlike most of the countries in Africa, the forestry department is a well-established and highly competent agency. And in addition, the goals of social forestry were established as government policy since 1976 (CSE 1985; DeRoy and Mathew 1988, 244). From this perspective, it must be understood that India had both the commitment to community forestry and the institutional capability for implementing the projects. It was also the Indian government, not the Bank, that largely demanded, designed, and implemented these project loans.

India's community forestry program most certainly has been controversial and the results mixed (Blair 1986; CSE 1985; DeRoy and Mathew 1988; Shiva 1991). These discussions need to be qualified, of course. People forestry projects financed by the Bank, as well as other funding agencies, existed throughout all of India. The successes of these programs, however, has varied considerably. Some were quite successful in assisting the poor, others were not. Still, experts have suggested that a significant portion of these programs tended to shift away from the ideals of community forestry. The program did have some spectacular successes. For example, millions of trees were planted. In the state of Gujarat alone in 1985, 200 million trees were grown, far beyond the original goal of 30 million (Blair 1986, 1317). Similarly successful programs could be found throughout India, especially with its farm forestry components.

India's community forestry program also had some disappointments. The most noticeable was the conversion of the community forestry goals of assisting the rural poor with fuelwood, fodder, and other needs to the production of wood by rural elites for commercial and industrial purposes in the urban areas (Blair 1986; CSE 1985; DeRoy and Mathew 1988; Shiva 1991). Prime, irrigated agricultural land often was used to grow tree crops at the expense of the food production. There was sometimes a decrease in rural employment since the tree production was less labor intensive than raising crops, and absentee landownership increased markedly. And with the widespread use of eucalyptus, few secondary benefits were derived from its production. Unlike the Bank's efforts in Africa, the tree growers of India tapped into lucrative pole and pulp markets in the cities. The boles of the eucalyptus trees are ideal for construction scaffolding, while this and other species were used in the production of pulp and rayon.

What is not fully understood is the degree of commitment by Indian foresters to the goals of people forestry, or the pressures placed upon the agency by external interests. Several experts have noted that although Indian Forest Service officials were pioneers in social forestry, they were also classical, traditional foresters trained to meet commercial and industrial demands. And bureaucratically, meeting and exceeding production goals were more important than analyzing impacts of projects. Consistent with the empirical and theoretical literature on the diffusion and adoption of innovations, wealthy adopters tended to have the social and economic resources to invest in new ideas, while the poor did not. The Indian foresters discovered that the wealthier farmers were more willing and better able to adopt this new innovation, so the foresters deliberately focused on the wealthier farmers. DeRoy and Mathew (1988) indicate that working with the rural communities and the poor was much more difficult than working with the landed elite (245–246). Mistrust of governmental officials, issues of tenure and equity, and the other costs associated with collective adoption processes (West 1983) helped to constrain the foresters and their efforts to promote community forestry among the groups it was intended to benefit.

While it is the government of India that deserves most of the credit and blame for the successes and failures of its community forestry program, the Bank's role needs to be discussed a bit more. By all accounts it appears the Bank played a passive role in its relationship with the Indian government. But that is exactly the point. As was expressed by several experts I interviewed, Bank officials did little to monitor the projects. The Bank should have realized the projects were going astray and should have insisted on some changes. But the Bank did not. As indicated earlier, the Bank has tended to lack serious interest in monitoring the performance of its projects.

Summary of the Findings

A review of the evidence just presented shows that a substantial portion of the World Bank's forestry loans during the period went to rural development forestry projects. This has been particularly true since fiscal year 1978, when the Bank called for such loans with the publication of its first forestry sector policy paper. From fiscal years 1978 to 1992, more than $1.1 billion, or about 41 percent of the total value of the Bank's forestry loans, was allocated to rural development. During the same period, traditional forestry projects composed about 42.4 percent, or about $1.2 billion, of the Bank's forestry lending. The remainder went to projects that had both traditional and rural development components.

Asia captured the lion's share, about half, of the Bank's rural development forestry lending. India alone accounted for $470 million, or slightly more then 41 percent of the Bank's total rural development forestry lending. Latin America, on the other hand, acquired only 21 percent of the total. Only four projects, totaling $242 million, went to Latin America and the Caribbean, beginning mostly in the 1990s. Although it had four times the number of projects as LAC, the continent of Africa captured only about 19 percent of the lending.

With its original forestry sector policy paper, the Bank was very successful in stimulating its work in rural development forestry. The Bank, however, has had difficulty in sustaining its lending activities in both traditional and rural development forestry. A former high-level Bank official in an interview revealed that there has been considerable disappointment in forestry as a means of moving the Bank's money (Brechin 1989). As an illustration, forestry loans as a percentage of the total lending, while nearly doubling over the 1978–1992 period, has still remained quite low. In fiscal year 1978 forestry commanded a 1.19 percent share of total lending, whereas by fiscal year 1992 it had increased to only 2.1 percent (Brechin 1994; World Bank annual reports).

India's significant role in rural development forestry provides a fine illustration of the coming together of the Bank's technology, its relationship with donors and recipients, the nature of forestry projects, and the rural development task. India was the leading borrower of Bank loans for rural development forestry purposes in part because the Bank can easily make loans to India. This was not just in forestry. India and Brazil have been the Bank's biggest customers. Rural development forestry loans were *in demand* by the Indian government, and it had the institutional capacity *to absorb* significant amounts of financing. India continues to have a well-established Forest Service. It has a fine professional reputation and has been a premier civil service position. The professional forestry journal, *The Indian Forester,* which is still published today, was founded in 1875. India is also a country with a vibrant economy that has allowed it to accept *large loans;* yet because of its large population size, and consequently low per capita gross national product, it qualifies for *IDA* (International Development Association) financing at concessional rates. All of India's rural development forestry projects were financed through IDA.

IDA financing was often crucial since rural development forestry usually has not compared well to other projects that generate more income, although India's case was largely an exception. As a result, neither the Bank nor borrowing countries were eager to promote further indebtedness for the purposes of rural development forestry. Few countries could meet the Bank's internal requirements for significant lending in forestry or match India's ability to pursue rural development forestry loans. The interaction between the Bank, its technology, its relationship with clients, and the nature of the forestry sector limited the Bank's forestry work, especially in rural development forestry. Latin America, however, is an exception. But because of its relatively vast forest resources, especially in Brazil, there was until recently little interest in Bank support for rural development forestry there.

The countries of Africa lacked both the desire and certainly the ability to invest heavily in forestry. Other needs were seen as paramount, and the African countries lacked the institutional capacity to absorb Bank loans. As a result, there was a scattering of relatively small loans to a large number of countries in Africa. The weak forestry institutions in Africa have severely limited the Bank's work there (Braatz 1985; Leach and Mearns 1988; World Bank 1978, 1986a) Simply put, demand for forestry loans during the 1978–1992 period was spotty at best. Forestry was not a very good vehicle for moving the Bank's money in the 1970s and 1980s and was not given a great deal of support within the organization. With a couple of exceptions, the loans were, by Bank standards, small. In addition, economic and financial purists generally were skeptical of the economic appraisals, especially of rural development forestry. Studies undertaken by the Bank indicated that social forestry projects could provide economic rates of return of 20% to 30% (see World Bank 1986a). Still, others inside and out of the Bank remained skeptical of the studies or of the promise of community forestry itself. Rural development forestry projects have not stacked up well against other projects promoting economic development. Forestry projects generally do not generate cash from exports, have long payback periods, and have benefits that are difficult to quantify (Braatz 1985; Rich 1986; World Bank 1978, 1986a). For these very reasons, demand for forestry loans from client countries was limited. Government officials, too, found these same faults. In general, there was little appreciation for and understanding of the value of rural development forestry in the process of economic development (McGaughey 1986).

These reasons help to explain why the Bank did not do more in forestry during the 1980s, especially in rural development. But what of the work it did do? Our review of the appropriateness of the Bank's rural development forestry program reveals that slightly less than 27 percent of the value of the Bank's loans and credits went to those countries in greatest need of this type of assistance (see Table 4.1).

The Bank's effectiveness in meeting technical objectives of its community forestry program was generally very poor. Most of the projects labeled as this type of forestry essentially became new types of industrial forestry, with the benefits accruing to urban dwellers, the rural elite, or government. The Bank tended to perform best when its role was limited to investing additional financial resources in established and proven projects that were constrained by the lack of capital. The institutional competence of the recipient government and its commitment to the goals of assisting the rural poor were key ingredients in the success of Bank-sponsored forestry projects. The Bank could also play an important role as a catalyst in promoting change in the arena of economic development, as shown by the impact of the Bank's 1978 forestry sector policy paper. The most recent forestry sector policy paper (published in 1991) continues the Bank's emphasis on rural development while, for political reasons, purposely reducing its support of logging. As the Bank continues to write forestry sector policy papers, it will be interesting to see if it can improve its lending and performance in rural development forestry. This analysis indicates it will be highly difficult for the Bank to do so unless it somehow radically alters its approach.

FAO's Forestry Program, 1980–1990

The most distinctive features of FAO Forestry's work were its volume and distribution of activities. From 1980 to 1990, FAO was involved in 690 projects in 113 countries (Brechin 1994). This involvement is far more than the World Bank and CARE combined. The same holds for its work in rural development forestry. This record reflects FAO Forestry's international service agency character, its legitimacy as the world's forestry leader, and the fact that its assistance program is based on grants instead of loans, funded largely by a single multinational lender, UNDP.

Figure 4.4 (a and b) gives the breakdown of FAO Forestry's work by type of forestry project. Overall for the period, FAO's forestry projects were valued at nearly $560 million. Just fewer than half of the projects (331), and 45.5 percent of the total dollar value, or $255 million, were devoted to traditional forestry projects. Another $142 million, or about 25 percent of the total value, went to rural development forestry projects. Almost 29 percent, however, went to mixed projects. This percentage is much higher than the Bank's 15.4 percent. As we shall see, CARE USA, on the other hand, had no mixed projects since its forestry program was geared exclusively to rural communities, not industry.

Most of FAO's projects (273) and resources ($225 million, or about 40% of the overall total) went to Africa. Asia was next, then Latin America, and then EMENA (see fig. 4.5).

FAO Forestry has operated in more countries and regions than the other two organizations. More than the Bank, FAO distributed its projects evenly and according to need. As a comparison, India, which figured prominently in the Bank's forestry activities, had only three FAO projects, valued at $1.4 million, or just less than 10 percent of its rural development forestry total (see Brechin 1994).

Unique to FAO Forestry were its global projects. During the period in question, FAO sponsored 20 interregional projects at some $18.5 million, though these made up only 3.3 percent of the total. Neither the Bank nor CARE USA had projects of such scale. These activities reflected FAO Forestry's distinctive quality as the world's forestry agency.

If one breaks down FAO Forestry projects by funding source, UNDP underwrote 336 of the 690 projects (57%) at nearly $321 million. Donor-member trust funds provided $220 million, or around 39% of the total expenditures. The remainder came from FAO's own small TCP fund, which still sponsored 196 (28%) of FAO Forestry's projects.

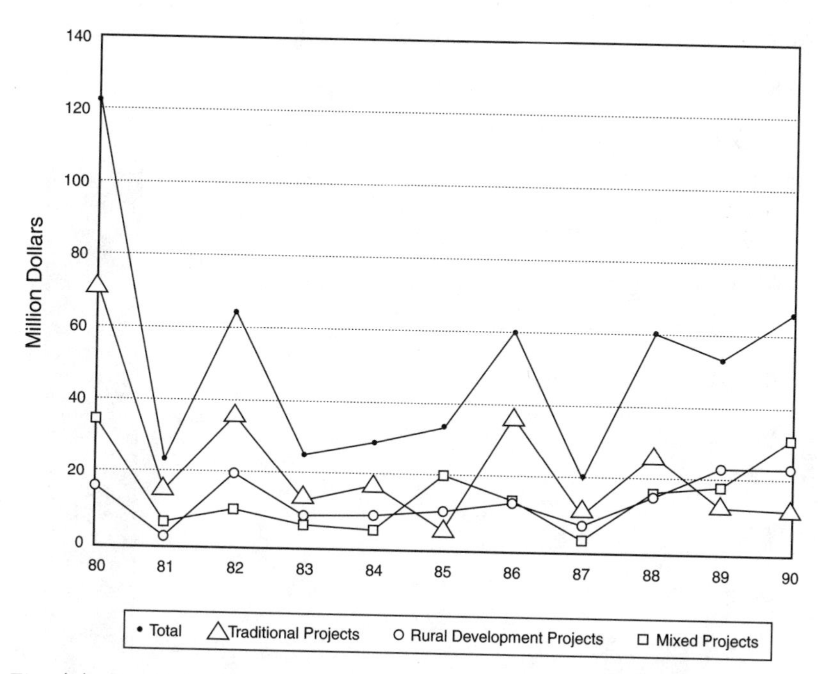

Fig. 4.4a. Total FAO Forestry Project Budgets by Type, 1980–1990 (Nominal Values)

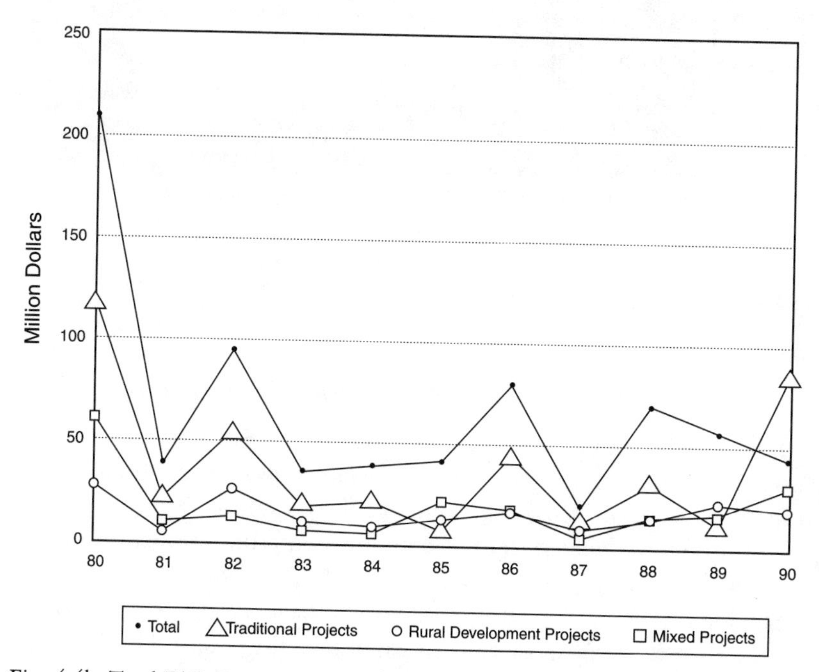

Fig. 4.4b. Total FAO Forestry Project Budgets by Type, 1980–1990 (1992 Dollars)

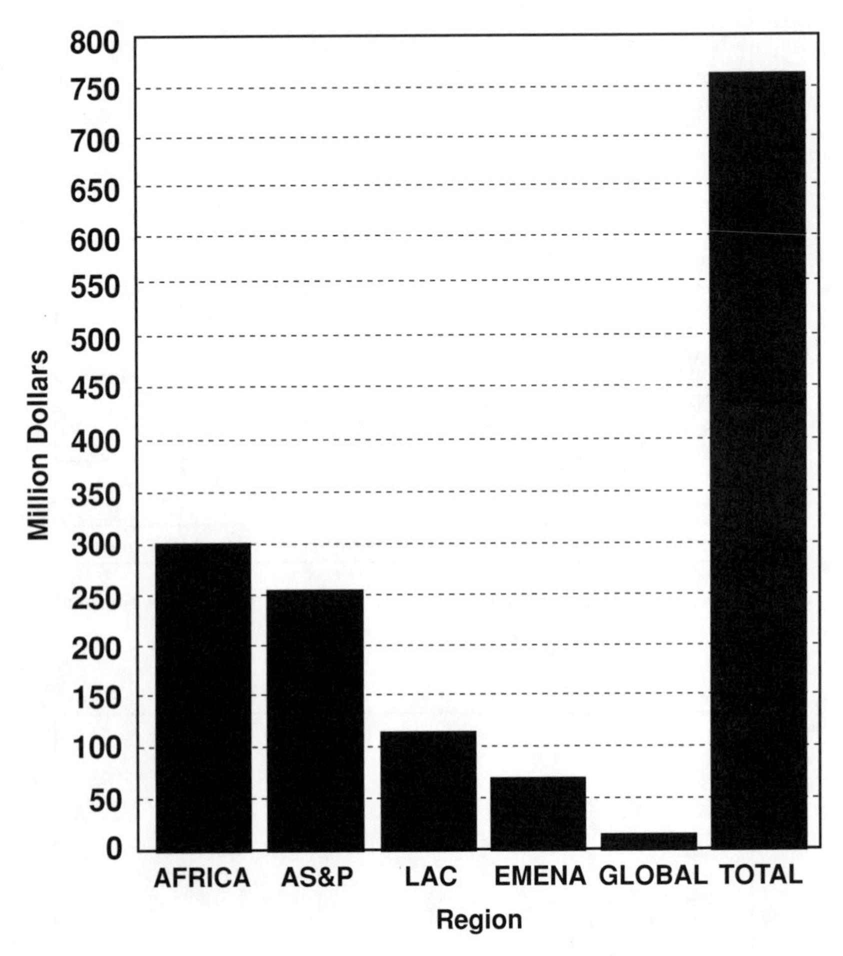

Fig. 4.5. Total FAO Forestry Budgets by Region, 1980–1990 (1992 Dollars)

FAO's Commitment to Rural Development Forestry

As seen in figure 4.4b, rural development forestry slowly increased through the 1980s with a sharp rise in the late 1980s. By 1990 some $185 million (see figure 4.6), or about a quarter of the value of FAO Forestry Department's forestry work, went for community and rural development projects. These figures confirm FAO Forestry's decisive movement into rural development forestry. Still, the percentage of the value of FAO community forestry is considerably less when compared with the Bank's. The reason for this low commitment is due, however, more to the constraining nature of FAO Forestry's task environment than to the character of its technology.

Within this new type of forestry, Senegal led FAO's list with 11 projects valued at $17.9 million, which is less than 13 percent of all FAO Forestry's rural development projects and only 3 percent of the overall forestry total. Cape Verde was second on the FAO list, with $14.9 million. There were 189 rural development forestry projects. Most (57%) were situated in Africa; but unlike the Bank, FAO Forestry did not focus on one or two countries for its rural development program. Instead, it was able to spread its work across a number of countries and continents.

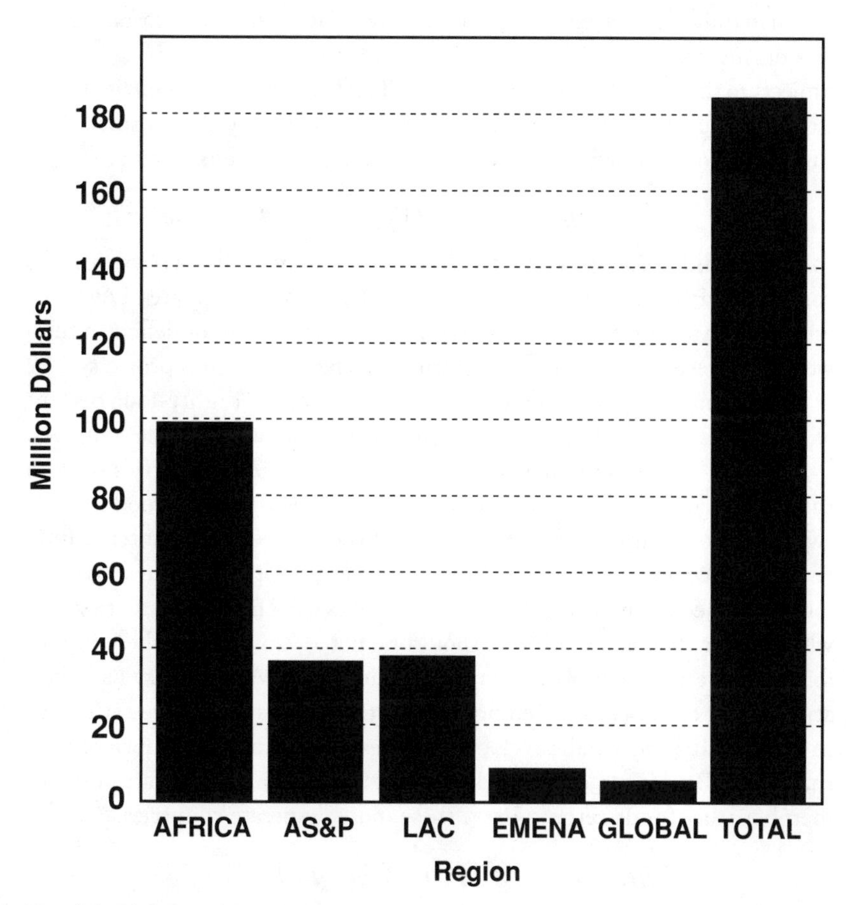

Fig. 4.6. FAO Rural Development Forestry Budgets, 1980–1990 (1992 Dollars)

The regional distribution of the total value of community and rural development forestry projects for the period was 19.6 percent for Asia, 53.9 percent for Africa, 19.8 percent for LAC, 4.3 percent for EMENA, and 2.4 percent for global and interregional projects.

The trust funds were the principal funding source for community and rural development forestry. FAO Forestry's largest funder of field projects, UNDP, allocated only 17 percent of its resources to rural development projects. Trust funds from donor governments, however, allocated over 41 percent of their financial support to these new types of projects. The smaller overall size of the trust funds and the relatively large portion of UNDP funding devoted to traditional forestry constitute the principle reason why FAO Forestry was less than fully committed to community and rural development forestry projects during the 1980s. The rapid rise of both trust fund support and mixed projects in the late 1980s was the result of TFAP, when donor countries began to fund national forestry plans. Just under a third (30%) of FAO's small TCP support in forestry had been allocated to rural development projects.

Appropriateness of FAO's Forestry Program

How did FAO do in allocating rural development forestry projects to countries with the greatest needs? Table 4.2, below, compares FAO Forestry's work to its own list of countries with acute scarcity or deficit in fuelwood.[5] As seen in the table, FAO during this period had 103 projects in 22 of the 25 countries facing serious fuelwood problems. Eighty-four percent of FAO Forestry's rural development projects were located in these countries.

Table 4.2 also compares the appropriateness of FAO Forestry's work to those countries with both acute scarcity and a dependence on wood for over two-thirds of their energy supply. In 10 of the 12 neediest countries, FAO Forestry executed 66 projects, for 47 percent of its rural development forestry projects. These figures are not as high as CARE USA's, but are somewhat higher than the Bank's. It should be noted as well that these figures for FAO Forestry are probably on the low side since FAO Forestry had a relatively large category of mixed projects. In these terms, FAO Forestry's work in community and rural development forestry was more appropriate than that of the Bank's. As elaborated later, this was largely the result of donor member trust funds zeroing in on these more desperate countries.

Effectiveness of FAO's Work in Forestry

The evaluation of FAO Forestry's performance is much more complex and muddled than the Bank's evaluation. Overall, sources gave FAO For-

Table 4.2
Appropriateness of FAO Forestry Department's Work in Rural Development Forestry: FAO Forestry Projects Compared to Countries with Acute Scarcity in Fuelwood
Fiscal Years 1980–1990

Countries	*$ Amount (millions)*		*No. of Projects**	*% Total Rural Development Forestry*	*% Total Forestry*
	Nominal	*Adjusted*			
Botswana	0.03	0.05	2		
Burkina Faso†	6.5	8.5	12		
Burundi	2.8	4.4	2		
Cape Verde	17.8	23.2	3		
Chad†	4.4	5.1	4		
Comoros	0.03	0.05	1		
Djibouti	0.43	0.5	2		
Ethiopia†	2.8	3.6	8		
Kenya†	1.36	1.86	5		
Lesotho	3.0	4.5	4		
Mali†	2.8	3.6	3		
Mauritania	8.3	10.4	9		
Mauritius					
Niger†	6.2	7.3	7		
Rwanda†					
Somalia†	5.2	5.9	4		
Sudan†	15.0	19.0	5		
Swaziland					
Afghanistan†	0.8	1.16	1		
Nepal†	14.5	20.36	9		
Bolivia	4.6	6.26	5		
El Salvador	2.7	4.0	4		
Haiti†	7.0	9.9	8		
Jamaica	0.1	0.17	1		
Peru	13.25	18.6	4		
	119.60	152.15	103	84.2	21.35‡
	66.56	86.28	66	46.9	11.9§

Source: Data compiled from Brechin, "Forestry Project Data for the World Bank, FAO, and CARE International" (1994, mimeograph).
*In an effort to be conservative, 34 mixed projects have been included in the totals. All of the countries except Botswana, Chad, Comoros, Kenya, and Mauritania had mixed projects. Afghanistan and Jamaica had only "mixed" projects.
†Indicates a country that depended on fuelwood for over two-thirds of its total primary energy consumption in early 1980s (see *Tropical Forestry Action Plan* (1985), p. 49).
‡Data in this row reflect FAO Forestry projects in 22 of the 25 countries listed in the table.
§Data in this row reflect FAO Forestry projects in 10 of the 12 countries listed in this table with the most acute fuelwood needs (i.e., those countries that used fuelwood for at least two-thirds of their energy needs).

estry highly variable marks in its community and rural development forestry activities. But, as in the discussion of the Bank's performance (and that of CARE USA to follow), careful consideration must be applied to the criteria of the evaluation. Most obvious in this regard was FAO Forestry's service-agency relationship to its recipient member governments.

Even as an executing agency, FAO Forestry's role was geared largely to assisting the recipient government in a project's implementation. The government and funder generally controlled the financial resources while the technical services were the responsibility of FAO. As a consequence, it was the interaction of the tripartite relationship—FAO, funding agency, and recipient agency—that seemed most crucial in determining the success of FAO Forestry's field projects. It was less FAO Forestry's technology or structure that affected its behavior and performance, as one might have thought. But before we go further into this assessment of projects, we should review FAO Forestry's Regular Programme.

Aside from its role in TFAP, FAO Forestry generally received praise for the activities of its Regular Programme. This was especially true for its efforts under the Forests, Trees and People Programme (FTPP), which somehow avoided the TFAP ruckus. In addition, FAO Forestry is much more than TFAP. Its publications, workshops, seminars, training courses and other educational and academic activities were considered superb and generally well executed. This was FAO Forestry's role as developer of the forestry intellect. It was in this capacity that FAO Forestry needs to be recognized as the intellectual innovator of the community forestry concept. (One needs to pause here, however, and consider the work of individual forestry pioneers in India, as well as the foresight of FAO Forestry's own Jack Westoby and Mike Arnold). Regardless, it was FAO Forestry that more fully developed, generalized, and diffused the technical concepts of community and rural development forestry. And as I indicated, FAO Forestry's role in the process was constrained until Sweden interceded.

It is important to note that the dimensions of FAO Forestry's relationship with its member governments were different with its Regular Field Programmes. In particular, the Regular Programme tended to be more of a simple and straightforward didactic relationship, as between able instructor and willing student. This educational service was optional, not the least bit compulsory. Obviously, this selection bias helps to maintain quality. It was also a simpler, bipartite, as opposed to a more complex tripartite, relationship, since funding agencies were not usually directly involved in Regular Programme activities. In addition, SIDA (Swedish International Development Agency) itself was unique as a relatively unpolitical development assistance organization. And perhaps more important, financial resources, and the control over their allocation, were not part of the Regular Programme agenda. All these factors seemed to contribute to the overall greater success of FAO Forestry's Regular Programme. In summary, the content (thought not the execution) of FAO Forestry's activities were largely shaped, demanded, or otherwise approved by the member governments. Within the Regular Programme, the professional technical skills of FAO Forestry's staff, that is, the organization's technology, were able to work largely in an unconstrained manner. Once given marching orders to proceed in a new direction by member nations, its day-to-day operations were rarely interfered with.

Not unexpectedly then, FAO Forestry's poorer technical performances were associated with its Field Programme activities. This was the result, already alluded to, of greater constraints placed upon FAO Forestry's technology as a result of its complex interactions with members of its more powerful task environment. As a consequence, the success of FAO Forestry's work in the field was determined largely by the interactions among FAO Forestry field personnel, the recipient government forestry officials, and the donor agency staff. If the interactions were appropriate and constructive, and in support of the project's technical aspects, then the likelihood for project success increased dramatically. If the interactions were conflictual or otherwise less supportive, then the likelihood for project success decreased dramatically. Within the tripartite relationship, FAO Forestry personnel had a weak hand to play and were compelled to play extremely well with the few cards they had. Here, the professional and diplomatic skills of the FAO Forestry personnel were essential. A number of informants mentioned that an excellent FAO Forestry staff can create a successful project out of a touchy situation, while a less skillful group could easily transform an otherwise favorable situation into a disastrous one. One of the more common scenarios was a clash of wills, style, or authority between FAO personnel and host-government officials, with the funding agency providing little in the way of guidance.

In comparison to the Bank, projects designed by FAO Forestry field staff tended to fare much better. Here, FAO Forestry's technology seemed to be very well matched to the task. And with the presence of FAO Forestry field personnel at the project site, the organization maintained an appropriate, although a temporary, decentralized structure for rural development project tasks. It was generally agreed that FAO Forestry headquarters staff provided first-rate technical advice to its field personnel. These facts tended to reemphasize the constraints that its task environment placed upon FAO Forestry's activities in project implementation.

Of course, not all rested on the shoulders of the FAO Forestry staff. The recipient governments themselves needed to be committed to the technical goals of the projects, that is, to community and rural development forestry, and to demonstrate a sincere willingness to assist the rural poor. Likewise, the donor agency had to be willing to be flexible and supportive of the needs of the executing agencies in their implementation of the project. The funding agency, especially donor countries, could also create the setting for better performance by establishing certain criteria necessary to enhance the project's technical objectives. For example, for the field project component of the FTPP and its predecessor, the Swedish government established some basic criteria for participation. These requirements were that (1) projects must be designed to assist the rural poor, (2) governments must be committed to the project's goals, (3) local people must be involved in the project, and (4) the institutional capacity must exist or be created to carry out the project (FAO 1985b, 34). This also allowed FAO Forestry staff to maintain its technical competence while shedding more of the delicate politics of placing conditions on another member as "the funder's requirement." When the tripartite factors come together in a positive fashion, as they generally did in Nepal and Peru, for example, projects can be very successful. Of course, the opposite was also true. The use of conditions on participation was less likely for UNDP (United Nations Development Programme) projects, since funding here tended to be controlled more by recipient countries.

Finally, we should consider FAO Forestry's target population. The character of the organization requires it to serve, not the poor people of the recipient government, but the recipient government itself. Often the work of FAO Forestry had to be translated into assisting the rural poor through an important intermediate step—the government agency. It could be easily argued that helping the poor and helping to develop the capacity to help the rural poor are not exactly the same thing. But it might also be argued that the work of FAO Forestry could be more substantial and sustainable than would otherwise have been the case if this type of work was not undertaken within the government. This is discussed again when we review the work of CARE USA.

Summary of the Findings

A review of the evidence shows that only a moderate portion (25%) of FAO Forestry's projects went to rural development forestry. Traditional forestry practices made up nearly 46 percent of FAO Forestry's field activities. The remainder were mixed field projects or those whose components were not discernable. Of particular interest was the fact that member donor countries through their voluntary trust funds were the principal sponsors of rural development forestry. Over 41 percent of their financial support and about 42 percent of their sponsored projects were forestry projects of this type. UNDP, the largest financial supporter of FAO Forestry field projects during the period, financed only 17 percent of FAO Forestry's rural development and about 28 percent of the total number of rural development forestry projects. UNDP's fairly meager support severely limited FAO Forestry's work in rural development. During the period, 83 percent of UNDP's support went to traditional and mixed forestry (Brechin 1994; see also Muthoo 1985, 1991).

Unlike the Bank, FAO Forestry projects were not dominated by a single country. Its projects were much more evenly distributed. Traditional forestry projects tended, however, to be concentrated in Asia, with most of the rural development forestry projects in Africa.

These findings highlight an unexpected conclusion, namely, that FAO Forestry, at least in its Field Programme, performed as a relatively neutral and flexible executing agency during the period. It attempted to meet the demands placed upon it by its donor and recipient member countries. Without significant resources of its own to devote to projects of its choosing, FAO Forestry's Field Programme *reflected* more the desires of its task environment than of its own will and vision. As a service agency, it assisted its member countries as they wished and as the limits of the professional knowledge of its staff allowed. In evaluating FAO Forestry's performance, it is important to remember that member-government forestry agencies and officials were the target populations. To meet the demands of community forestry, FAO Forestry radically altered its professional technology to incorporate this new knowledge as part of its character and organizational operations. While the organization adapted its technology to meet the new demands of its task and institutional environment, it was ironically the same task environment that ultimately limited its work in this new area.

Contrary to what many would have expected, FAO Forestry established a relatively appropriate community forestry program. Eighty-four percent of its rural development forestry activities took place in those countries were they were needed. FAO executed 66 projects in 10 of that 12 most crucial countries, with almost 47 percent of the value of its rural development forestry program. The appropriateness of FAO Forestry's work was largely the result of trust fund support from donor nations.

The effectiveness of FAO Forestry projects was highly variable. FAO Forestry's Regular Programme work in community and rural development forestry was considered superb, but usually a step or so further removed from directly assisting the rural poor by its mission in educating and advising government officials. At the same time, it can be argued that these activities are essential if forestry efforts to reach the rural poor are to be realized on a consistent and institutionalized basis.

While the rural poor may indeed have received attention from FAO Forestry's field projects, they were more likely to remain, at best, a target population secondary to the forestry agency itself. As noted, FAO Forestry field projects tended to vary more in their level of performance. The competence of FAO Forestry field staff, in general, was less consistent than that of its Regular Programme staff. In addition, performance was greatly hindered by the tripartite relationship among FAO, the donor agency, and recipient government. Project success depended largely upon the complex and uncertain interaction of the players within this arrangement.

CARE'S Forestry Program, 1976–1992

A number of interesting patterns emerge from a review of CARE's forestry program. Figure 4.7 illustrates the dramatic increase in the total value of forestry projects managed by CARE during the period in question. In fiscal year 1976, CARE's forestry budget was only $114,038, for the original projects in Niger and Guatemala. By fiscal year 1992, the value rose to slightly more than $18 million. Some $104.4 million was administered by CARE for forestry projects during the 17-year period (see Brechin 1994).

One can also see a comparable, although not unexpected, increase in the number of forestry projects sponsored by CARE. There were only a few projects under way in the 1970s and early 1980s. A rapid increase began in fiscal year 1982, with 25 ongoing projects by fiscal year 1987 and 42 in 1992. During this period, CARE sponsored 96 forestry projects. Some of this increase could be attributed to the CARE-USAID matching grant program, but most of CARE's increase in forestry activities came from increased interest from a variety of funding sources.

An analysis of the geographic distribution shows that most of the value

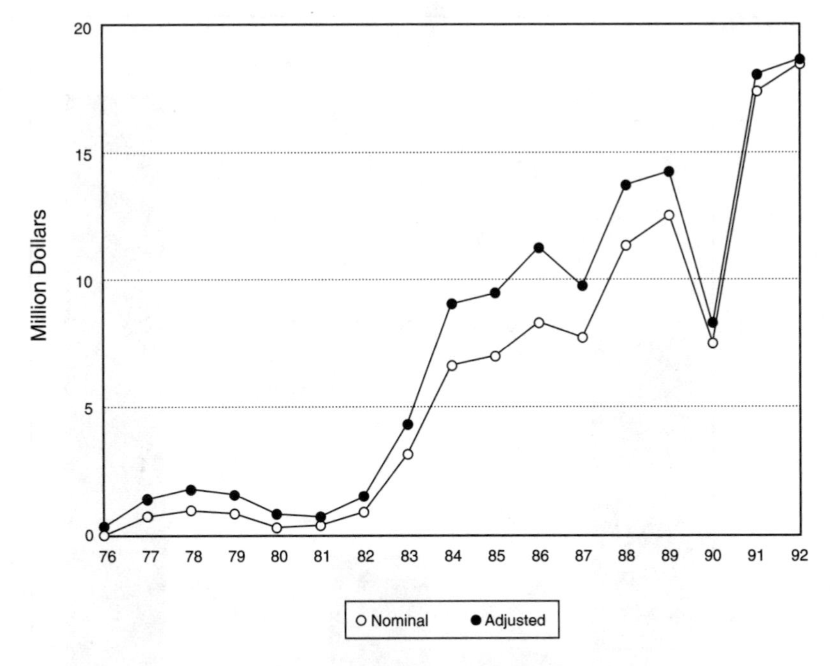

Fig. 4.7. Care Total Budgets in Forestry, 1976–1992 (Nominal and 1992 Adjusted Values)

($64.1 million, or 61%) of CARE's forestry projects (all for rural development) were located in Africa. LAC, Asia, and EMENA trailed (see fig. 4.8). Of the number of projects, 56 percent were located in Africa, while only 28 percent and 13.5 percent were in LAC and Asia, respectively.

As is evident, CARE concentrated its work in forestry in one region: Africa. The Bank tended to concentrate its efforts in rural development forestry too, though in South Asia, especially India and Nepal. For CARE, the country with the largest total forestry budget was Niger, with $12.4 million, or 11.9 percent of CARE's total forestry budget. Mali was next with $6.5 million, or 7.6 percent of the total. Uganda was third ($6.8 million) and Sudan fourth ($6.5 million). Within Latin America and the Caribbean, Ecuador had the most activities, with projects budgeted at about $6.2 million (5.9% of total). Haiti closely followed with $5.9 million (5.6% of the total). In Asia, CARE had projects in only three countries, Nepal led that region with $5.9 million in projects (5.6% of total). Afghanistan was second and Thailand third with $4.3 and $1.6 million in projects, respectively.

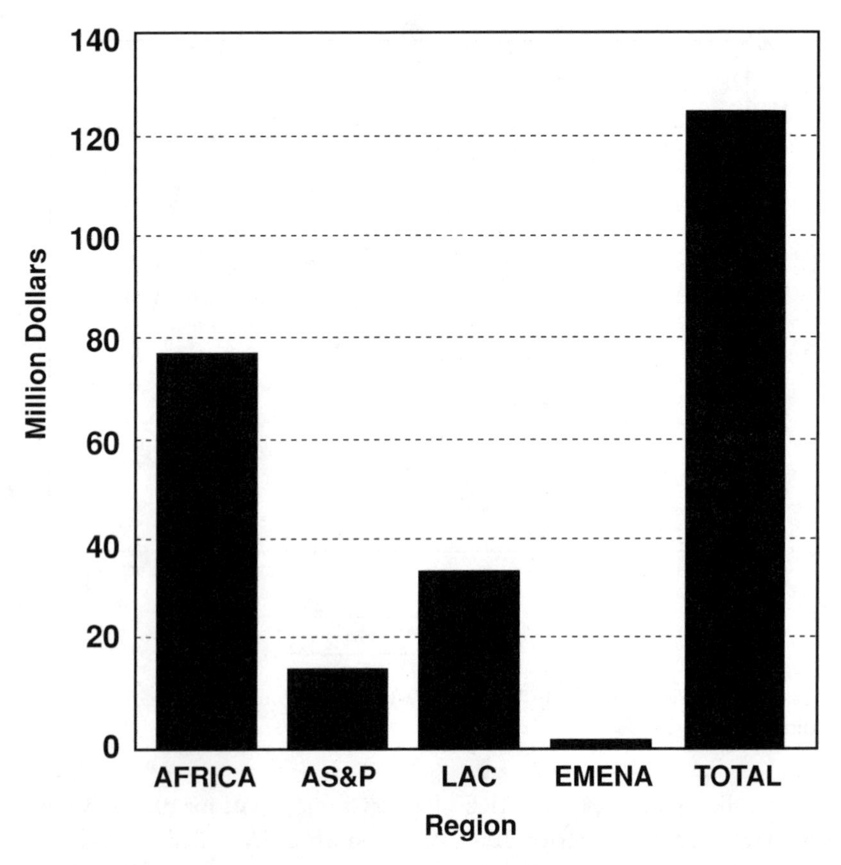

Fig. 4.8. Total CARE Forestry Budgets by Region, 1976–1992 (1992 Dollars)

CARE's Commitment to Rural Development Forestry

Since CARE's forestry program consisted only of rural development forestry activities, its commitment to rural development forestry could not have been higher. When comparing the value of rural development forestry projects against all CARE projects during the period, one sees a dramatic increase in both the overall yearly forestry project budgets and in the proportion of forestry to the overall field program expenditures. For example, the forestry budget for fiscal year 1976 was $114,000, or only 0.07 percent of CARE's work overseas. By 1987 the proportion was 2.8 percent, and in 1992 it was 7.4 percent (see Brechin 1989, 1994; CARE 1992). CARE often controlled project funds that were usually collected from a number of sources, including private contributions from CARE USA or CARE International members, recipient government contributions (usually in-kind) or third-party donations from such agencies as USAID. Increased funding also came from multilateral agencies such as the United Nation High Commissioner for Refugees.

The rapid increase in CARE's forestry program began in fiscal year 1982 (September 1981) when CARE signed a four-year, multimillion-dollar Renewable Natural Resources Matching Grant with USAID (CARE 1985c). This grant of $2.73 million was matched dollar for dollar by money raised by CARE's fund-raising drive known as "CARE for the Earth Campaign," an additional $1.2 million coming from recipient-government contributions (CARE 1985c). In all, over $6.6 million was made available from fiscal years 1982 to 1985 for 10 community forestry projects in seven countries in Africa and Latin America (CARE 1985c). Although USAID's portion of the Matching Grant was important, it was relatively small compared to the total. Only about 14 percent of the financial value ($2.7 million out of $19.1 million) of CARE's forestry program for those years was financed by the government (see Brechin 1989: Annex Table 5.3; CARE 1985c). USAID's support of CARE's forestry program continued in 1985 (fiscal year 1986) with the signing of a new multimillion-dollar "Partnership Matching Grant."

Appropriateness of CARE's Forestry Program

Table 4.3 shows the results of CARE's work in those countries with serious need. CARE had 58 forestry projects in 14 of the 25 countries listed. Overall, $71.6 million, or about 69 percent of the financial value of CARE's forestry projects, were located in these extremely impoverished countries.

If we focus on those countries with the most acute needs, CARE had 52 projects in 11 of the 12 countries (Table 4.3). Those 11 countries received $64.2 million, or 61.5 percent of CARE-managed project funds. Significantly, the nature of CARE's work changed little when we focus on only those countries with the greatest need. Again, this reflects CARE's character of serving mostly the poor of the poorest nations.

Effectiveness of CARE's Forestry Program

CARE's work in rural development forestry was for the most part well respected. Its projects tended to be appropriately designed, efficiently run, and geared to the intended beneficiaries of community forestry—the rural poor. In addition, the projects most often started simply and on a small scale. Projects that seemed to work were usually allowed to evolve to meet

Table 4.3
Appropriateness of CARE USA's Work in Rural Development Forestry:
CARE USA Projects Compared to Countries with Acute Scarcity in Fuelwood
Fiscal Years 1976–1992

Countries	*$ Amount (millions)**		*No. of Projects*	*% Total Rural Development Forestry*	*% Total Forestry*
	Nominal	*Adjusted*			
Botswana					
Burkina Faso†					
Burundi					
Cape Verde					
Chad†	4.9	6.9	4		
Comoros	3.3	3.7	2		
Djibouti					
Ethiopia†	4.6	5.0	4		
Kenya†	1.9	2.4	2		
Lesotho	1.4	1.7	2		
Mali†	8.0	8.8	6		
Mauritania					
Mauritius					
Niger†	12.4	14.8	14		
Rwanda†	5.6	6.4	3		
Somalia†	4.1	5.3	3		
Sudan†	6.6	8.2	4		
Swaziland					
Afghanistan†	4.3	4.3	1		
Nepal†	5.9	6.6	6		
Bolivia	2.7	3.0	4		
El Salvador					
Haiti†	5.9	6.8	3		
Jamaica					
Peru					
	71.6	83.9	58	68.6	68.6‡
	64.2	75.5	52	61.5	61.5§

Source: Data compiled from Brechin, "Forestry Project Data for the World Bank, FAO, and CARE International" (1994, mimeograph).
*Dollar figures are nominal and adjusted to 1992 dollars.
†Indicates a country that depended on fuelwood for over two-thirds of its total primary energy consumption in early 1980s (see *Tropical Forestry Action Plan* (1985), 49).
‡Data in this row reflect CARE USA's forestry projects in 14 of the 25 countries listed in the table.
§Data in this row reflect CARE USA's forestry projects in 11 of the 12 countries listed in this table with the most acute fuelwood needs (i.e., those countries that used fuelwood for at least two-thirds of their energy needs).

newly identified needs. Less-successful projects, like the effort in Egypt, were generally quickly dismantled, unlike the more politically driven projects of FAO, or the investment-driven ones of the World Bank. To help maintain its focus on performance, technical support and monitoring and evaluation efforts became increasingly sophisticated and formalized.

Most of CARE's work was directed to small farmers who owned or had access to land. Consequently, some questions were raised about CARE's ability to reach the landless poor and women. When compared with more grassroots-oriented, bottom-up PVOs like World Neighbors or Oxfam, CARE generally was viewed as much less participatory, or even heavy-handed, in its approach to communities. Consequently, some of the critical social dimensions of its projects, like tree tenure issues, were not worked out as well beforehand as they could have been. Still, CARE projects did not have problems of conversion of benefits from the rural poor to the government, urban dwellers, or rural elite. Finally, CARE's overall fine performance was tempered significantly by the limitedness of its projects. Many of CARE's forestry projects in the 1980s were small and isolated.

Perhaps most impressive of CARE's work was its ability to perform well in countries that are not particularly sensitive to bottom-up assistance for the rural poor. Here CARE's strengths of working more or less independently of government agencies and directly with the rural poor were essential. For the record, CARE must work *with* a government agency, but not for one. This is not to suggest that CARE's work was not shaped or otherwise affected by the actions of the host governments. It has, however, maintained almost complete control over its projects' resources, which has allowed it to concentrate on performance. CARE's agroforestry work in Haiti and Niger provided fine illustrations of this ability.

The Haitian government has a long and tragic history of indifference or even hostility to its rural population. The CARE project in Haiti, which was a subcomponent of a larger USAID project known as the Agroforestry Outreach Project, ran from 1982 to 1989. CARE ran decentralized nurseries. It also administered an effective and intensive extension program which, during its height, had 91 field staff, including 6 expatriate managers, far beyond the "average" community forestry project (Weber 1985, 8). From 1982 to 1986 the entire project, including the additional components, supplied 27 million trees to 110,000 farmers, far exceeding the original project objectives (Leach and Mearns 1988; Winterbottom and Hazlewood 1987, 108). This was accomplished without any government assistance.[6]

An important observation from the Haiti Agroforestry Outreach project had to do with the consequences resulting from a number of organizations working together. Although CARE's agroforestry component in Haiti was actually quite small, the collective effort sponsored by USAID was quite substantial, as the numbers above indicate. This helped generate interest in better utilizing PVOs to implement large forestry projects (see Conway 1987; Leach and Mearns 1988).

CARE's projects in Niger provided another example of its ability to work in unsupportive settings. Following its French colonial influence, Niger maintained a strong top-down approach to government management. This approach, combined with limited resources, stifled local-level initiatives (see Brechin and West 1982; Thomson 1977, 1981). CARE's ability to work with greater independence from national agencies and directly with local people created the essentials for more community-based forestry activities. It was able to provide the essential resources that were needed to make a project successful. The appropriateness of the Majjia Valley windbreaks design, and the visibility of their impact, especially through an increase in crop production, helped to generate greater enthusiasm and support for the project. The success in one village attracted farmers from neighboring villages who wanted to know when the project was going to come to their area. CARE could not, however, take all the credit. The area's district forester, a former agricultural extension agent, was an exceptional individual who maintained relatively good relations with the farmers. As CARE quickly found out, it was difficult to develop forestry innovations in areas where the forestry agents had already "poisoned" the forestry setting. CARE learned quickly and chose to work only in areas where the forestry official was well liked and competent (Brechin 1989; Brechin and West 1982; Delehanty, Hoskins, and Thomson 1985; Leach and Mearns 1988).

CARE could also assume the role for advocate of rural people as it attempted to protect the integrity of its projects. With the maturing and trial harvesting of the older sections of the Majjia windbreak in the mid-1980s, CARE had to negotiate firmly with the government to assure that the local people would receive sufficient benefits from the harvesting (Brechin 1989; Delehanty, Hoskins, Thomson 1985) CARE reasoned—no doubt, correctly—that if the government left insufficient benefits to the rural villagers, the latter would have become disillusioned and less supportive of the project. But critics claimed—also correctly—that CARE should have negotiated this very important issue with the government and the communities much earlier.

CARE's success was not limited to those countries where governments were unsupportive of bottom-up development activities. In 1984, CARE sponsored successful and extremely efficient village forestry projects in Uganda with the full assistance of the government's forestry department. Typical of most of CARE's projects, it began simply and small and evolved in directions that seemed appropriate to the needs of the local people. The project shifted from just producing free seedlings for farmers to more extensive work and training in agroforestry practices. It also shifted from the production of only a few limited species to more multipurpose species. From 1984 through 1986 alone, the project had in operation 274 nurseries, producing 10 million seedings, at 11 cents per seedling. This cost was the lowest of any USAID-funded project in Africa and the lowest for CARE worldwide at that time. It also gave underutilized government foresters a job to do and the resources to do it. There was great job satisfaction and high morale among the government staff (Tapp, Resch, Buck, and Ntiru 1986).

What makes the Uganda project unique was its success given the country's violent civil war at that time. CARE's expatriate staff were evacuated from September 1985 to February 1986. Although the fall 1985 and spring 1986 plantings were affected by the fighting, the Ugandan foresters were able to keep a number of the nurseries operating. The foresters also went to extraordinary lengths to protect project materials and equipment from confiscation or destruction during the war (Brechin 1989; Tapp, Resch, Buck, and Ntiru 1986). This allowed the project to continue later in 1986 without much difficulty.

In spite of CARE's ability to achieve the technical objectives of its projects, and its general success as a project innovator, its projects were often limited in their scope and impact. In this sense, they could be viewed as "micro" solutions to "macro" problems. Some of the project activities of FAO and the Bank had the same failings. But with CARE the failing were even more extreme and more frequent. Its mandate was to work at the local level, directly with rural communities. In fact, CARE's ability to work more or less independently of the national government as a PVO equally limited its legitimacy to work with the government on national issues. Because of their international public status and related powers, the Bank, through the use of its structural adjustment and sector lending, and FAO, through its recognized legitimacy, had considerable influence in shaping national policy issues and bureaucratic behavior, much more than CARE did. CARE, as well, was limited by its technology and character to work only in the poorest of the poor countries. Partly in consequence, CARE did not have the same geographic scope nor depth of financial resources as the other organizations did. It did, however, generally do a good job of reaching the poor.

This is not to suggest that CARE was ineffectual. It was, rather, merely limited in its effectiveness. This is true for the other two organizations as well. CARE was a superb innovator and frequently executed well on the ground. CARE did best when its technical accomplishments were noticed and promoted by other organizations with other strengths, especially political ones.

Summary of the Findings

CARE became systematically involved in community forestry in 1974, with the first substantial investments coming in 1975 (fiscal year 1976) when it approved the Majjia Valley windbreak proposal.[7] Almost simultaneously, CARE supported a reforestation project in the highlands of Guatemala. In both cases CARE was approached by government field personnel and was able to move quickly, providing resources within a few months, to support the projects in a timely fashion. With CARE's continued support, the scale of operations was quickly increased.

Essential to CARE's responsiveness to both projects was the presence of its own field personnel at its established country missions. This cannot be overemphasized. In short, CARE's fieldwork was program based and not simply project based. CARE attempted to develop long-term programs through the implementation of a number of projects among a number of appropriate sectors, rather than act simply as a temporary presence associated with a single project. Consequently, it generally established what were essentially, for programming purposes, permanent missions, and it was therefore in a position to act relatively quickly if new project ideas or needs arose. Historically, CARE tended to introduce feeding programs first in new missions and then develop new programs in different sectors as new needs were identified.

The projects in Niger and Guatemala provide classic examples of CARE's ability to expand into new areas by using essentially bottom-up actions, simply by its presence and willingness to respond to new initiatives. CARE's accessibility, established operations, and reputations within the countries allowed government field personnel (with eventually host-government approval) to approach CARE as a source of support for forestry projects.

CARE's total commitment to community forestry is obviously noteworthy. It had no commitment to forestry in general, only community forestry. Because of its decentralized decision-making, self-help community development approach, CARE was "ready-made" to support community forestry. Between 1978 and 1992, forestry within CARE had increased steadily from 0.07 percent of its overseas work to 7.4 percent. Much of CARE's increased involvement in forestry projects was linked to its ability to attract private donors to support conservation-related causes. In addition, with the creation of a permanent Renewable Natural Resources staff position at CARE USA headquarters in 1981, CARE was able to generate funding support through the development of grant proposals. CARE's multimillion-dollar matching grants with USAID provide the most notable case. CARE continued its centralized support for its forestry program by pursuing further funding agreements with more multilateral agencies (CARE 1988, 1994 interviews). External support certainly helped CARE expand its work in community forestry, but CARE was not completely dependent upon it nor were its specific projects controlled by external funders. CARE's successful experimentation in community forestry was largely rewarded with very flexible grants and donations from others to let CARE continue its pioneering work.

The appropriateness of CARE's program in community forestry is also noteworthy. As shown above, about 69 percent of CARE's work in community forestry was located in countries with an acute scarcity in fuelwood, with 61.5 percent of CARE's total expenditures in forestry located in the countries that needed the help most. Most of CARE's expenditures (61%) in forestry were-centered in Africa, where the deforestation and fuelwood shortages continue to have a severe impact on the rural poor.

The appropriateness of CARE's program was not by chance. CARE's mission and character is geared toward serving the poor in the poorest countries of the world. To put it more bluntly, CARE cannot expect to receive large sums of private donations for its work if it operates in countries where a dramatic need is lacking. Conversely, less-desperate and more-capable countries are less receptive to organizations like CARE because of their insistence on controlling the projects and its resources. With CARE's presence in these countries, and with its decentralized decision-making core technology, it tended to design and implement programs and projects that were reasonably appropriate for local-level development.

The effectiveness of CARE's community forestry program also was the result of its flexible, decentralized decision-making technology and community development approach, decentralized structure, and less-constraining relations with the host government. In sum, CARE achieved the technical objectives of community forestry because there was a good match between the rural development task and its internal and external organizational elements.

While its performance was high, its overall impact was often limited. Its projects were usually small and remote. It could create rational projects, but ones frequently surrounded by larger irrationalities. Its projects may be the beginning of community change and development, but continued success requires a different kind of effort. CARE could only offer small-scale solutions to what are generally large-scale problems. And its very ability to achieve technical success on the ground limited its impact at higher levels. When it did have a broader impact, it was in partnership with other organizations with different strengths.

Chapter 5

Examining Organizational Behavior

When we compare the World Bank, FAO Forestry, and CARE, we find that there are great differences in the structure, technology, and task environment of each organization, yet no difference is more striking than in organizational character. As was noted in the Introduction, the character of an organization is the product of its history and is shaped by its early interaction with powerful forces from within and without. Certain values, approaches, and responses become institutionalized, leaving marks on an organization so deep and fundamental that they are scarcely recognized and seldom questioned.

Comparing Organizational Character

From this study, it is obvious that the organizational character of the three organizations under discussion has had a tremendous influence on their work and performance in community and rural development forestry. Table 5.1 summarizes the differences.

The World Bank, for example, is a financial institution. It was created, along with its twin sister, the International Monetary Fund, to bring financial order out of chaos in the world of international finance. It was shaped early on by its close ties to the world's financial markets where it sells its securities and borrows money to raise the capital it uses for loans. As a Western capitalist institution, it operates on the firm premise that economic development is best achieved through the wise investment of scare capital to spur economic growth. But as a bank, it does not give money away. To remain professionally credible, it must make economically sound loans. To remain vital, however, it must move its money efficiently by signing loan agreements with recipient governments that meet the Bank's needs as much

Table 5.1
Comparison of Organizational Character

World Bank	Financial institution moving money for development
FAO Forestry	Service agency (for national governments) possessing diplomatic and forestry professionalism
CARE USA	Private humanitarian development organization helping the poor of the poorest

Note: Character is the product of historical forces.

as the Bank meets the governments' needs. Consequently, as a bank, it must invest in programs and countries where it can move money—not necessarily those countries most in need, although it often strives to do both. For a number of reasons already discussed, forestry in general has not been a good vehicle for moving the Bank's money. This is especially true of community and rural development forestry, with the exception of India and, more recently, Brazil. The returns from community and rural development forestry may make economic sense as far as general benefits and costs are concerned; these types of projects, however, make less financial sense as far as cash generation is concerned. Ministry officials of the developing countries frequently overlook community forestry projects and prefer instead those that can generate the hard currency they desperately need. The organizational character of the Bank, operating through its structure, interactions with the environment, but particularly its investment core technology, has deeply shaped its behavior and performance in community and rural development forestry.

FAO Forestry's character, on the other hand, has been shaped by its configuration as a specialized service organization. As a unit within FAO, the Forestry Department was created after World War II as an international governmental organization to provide technical expertise in forestry for its members. Forestry was seen as one piece of the puzzle in achieving greater economic prosperity throughout the world. As an agency of the international community, it exists to serve the technical wishes of its members, governed largely by a democratic process (i.e., one country, one vote). As a service agency with no independent resources of its own, it depends upon its members for political support, program guidance, and financial resources. And here lies the rub. Because of FAO's position, it must walk a very thin tightrope. It is a democratic organization whose political legitimacy is in the hands of the numerous developing countries, while its economic viability rests in the hands of a few wealthy nations. This complication places

FAO (and most other UN specialized agencies) in the extremely awkward position of having to please both camps—a reality that provides a particularly keen set of inherent strengths and weaknesses.

It has the legitimacy of being the world's forestry agency but is constrained by its lack of independent power from which that legitimacy is derived. Community and rural development forestry became a part of FAO Forestry's technical package because it was a *new type* of forestry demanded by some of the members at a time when that type of forestry also happened to be gaining currency on the international political stage. Its specific activities in community and rural development forestry are, however, negotiated ones. Its successes and failures depend mostly on the willingness of member governments to involve themselves. Although it does attempt to influence events through publications and the diplomacy of conferences, FAO Forestry still must first have the consensus of members to proceed in new policy directions. FAO must follow before it can lead.

Finally, the character of CARE USA has been shaped by a much different set of forces. CARE emerged from the collective desire of a number of private U.S.-based humanitarian organizations to more efficiently distribute their relief supplies to the European survivors of World War II. Shortly after its creation, however, it took on a life of its own, but it was unable to completely shake off the stamp of its birth. It has remained a PVO dedicated to the relief of the poor and unfortunate, with little regard to political, religious, or social origin. Although it expanded its work to include development activities within its relief efforts, CARE has maintained, more or less, its focus on poor people and communities. It works with governments, but not through them. It often controls its own resources and project implementation. This was particularly true of CARE's work in forestry. Most of the external financial resources it received came from grants with very few strings attached. This gave CARE tremendous flexibility with its forestry field projects. In addition, in an attempt to serve its target population, CARE tried to gear its programs and activities around identified needs. Decentralized structures, such as country missions, along with their decision-making authority, allow for more flexible planning and implementation in order to be fairly responsive to a variety of (highly uncertain) national- and community-level requirements. CARE, as a private, humanitarian organization, depending upon the generosity of others, could only survive in the rough-and-tumble business of development through efficient and effective administration of its resources directed to poor people in the poorest countries.

Community and rural development forestry was added to CARE's repertoire, not because it was a new type of forestry, as with FAO, or because it was a new vehicle for moving money, in response to an administrator's vision, as with the Bank. Rather, community and rural development forestry was added because it was needed by poor communities in poor countries and was seen by the world community as an appropriate rural development program to pursue. CARE is an efficient, hardnosed development organization that works hard to maintain the quality of its programs through intensive staffing and through control over project activities. As a result, CARE has, generally speaking, well been equipped to provide community forestry projects in the countries where they have been needed most.

Concepts behind the Behavior: A Review of the Findings

Reflecting on Table 5.2, below, I restate the conceptual explanations behind organizational behavior, beginning with the World Bank.

The World Bank

From our review of the Bank's work in forestry during the 1980s, its core technology, characterized simply as investing in development through project activities, has actually been fairly constrained and more inflexible than I would have expected. In particular, the appraisal process of determining the appropriate economic and financial analyses has caused the Bank to design individual projects in a rather inflexible top-down manner. This "blueprint" approach, which is most likely a product of the Bank's historical focus on physical infrastructure, is also endemic to the process of calculating investments. Since various investment return calculations need to be made, a number of constants or assumptions have to be incorporated into the project. Consequently, this approach causes projects to be rigid because changes in design would likely change the calculations and thus the expected economic or financial returns and the nature of the signed agreement. This rigid design process is much more appropriate for physical infrastructure projects than for those tasks that have more dynamic social interactions and dimensions, like community and rural development forestry.

I discovered from this analysis a more precise understanding of the Bank's task environment. Although the Bank remains a powerful organization, its relationships with client governments have often shaped its project activities. Many governments depend upon the Bank as an important source of much-needed capital. But the Bank needs loan agreements to move

Table 5.2
Conceptual Explanations for Organizational Behavior:
Differing Organizational Technologies, Structures, and Task Environments

World Bank	•Constraining, inflexible, top-down technology •Delimiting centralized structure •Influential input task environment and empowering and limiting dependencies with output task environment
FAO Forestry	•Politically flexible, negotiated technology •Supportive dual structure with constraints •Controlling input and output task environments (both constraining and supportive)
CARE USA	•Flexible, decentralized, decision-making technology •Supportive decentralized structure with community development approach •Supportive input task environment owing to performance and organizational character •Supportive output task environment through selection bias

its money. As a result, even through there is little doubt that loans will be made, government officials usually have at least the right to determine what kind of loans will be made. A few governments such as Brazil and India, for example, can dominate their relationship with the Bank given their status as large borrowers.

Although it is certainly understandable that client governments would want to control the form of their indebtedness, the Bank's lending process, however, tends to bias the borrower's project selection away from those that are not well understood or that do not generate the hard currency needed to pay back the loan. In addition, few governments seem willing to add to their debt to assist politically "marginal" rural people without at least some direct financial return. Community and rural development forestry projects, as we have seen, tend to have positive economic, but sparse financial, returns, which makes this kind of investment a hard sell to client governments. The majority of loans for community and rural development forestry have been only through IDA at concessional rates. These and other issues of task environment have had important impacts on the Bank's lending for purposes of community and rural development forestry.

FAO Forestry

One might expect that the traditional forestry profession would be FAO Forestry's core technology. This profession is technically specialized, narrow, and value laden (Kaufman 1960; Ness and Brechin 1988). Traditional forestry involves the management of *trees* and forests for industrial and other commercial purposes. Community and rural development forestry, on the other hand, focuses on the well-being of *people* through greater personal utilization of trees for people's needs. One might think that at FAO these two types of forestry would be inherently at odds with each other.

Instead of a technically narrow traditional forestry technology, FAO Forestry has had a broad, politically negotiable one. It is a technology that has included a wide-ranging view of the relationships of forestry to economic and social development in the world's rural areas—much more than simply the management of trees for industrial development. This broader, more inclusive approach to forestry has reflected the growing demands of FAO Forestry's larger political and institutional environments as well as its more narrow task environment. As a service agency, FAO Forestry has had to adjust its professional services to meet the changing demands of its member governments and the larger cultural shifts that influence the agenda of economic development.

It must be noted that FAO Forestry has by no means abandoned its traditional forestry perspective. Rather, it has only supplemented it. But in the process of doing so, forestry within FAO has managed to integrate some concepts that speak to broader social issues and missions of forestry. This accretion provides insights into how international organizations change.[1] Finally, I hypothesized that FAO Forestry had a constraining, centralized organizational structure. Instead it has both centralized and decentralized organizational structures. The dual structures reflect the near-autonomy state of its Regular and Field Programmes. Its decentralized structure fluctuates along with its Field Programme, which is a collection of sporadic and individual projects in member countries. Consequently, it is a structure that has function but not continuity, except in Rome, FAO's headquarters.

CARE USA

The nature of organizational environment is important to understanding CARE. As with the other organizations, for CARE it is important to distinguish between input and output task environments. CARE has generally had a supportive input task environment. Private individuals and organizations along with government agencies and others have been willing to support CARE's forestry program. Similarly, CARE's output task environment has also been generally supportive. This support, however, has been the result of a selection bias, since CARE has only worked in geographic locations where it believed it could perform well. This has helped create a close link between the organization's inputs and outputs.

A Conceptual Discussion of Organizational Performance

This discussion leads us to a consideration of the two specific possibilities presented originally in Ness and Brechin (1988), where we proposed that international governmental organizations (IGOs) have been more effective than international nongovernmental organizations (INGOs) in generating world consensus around development and welfare issues (272). INGOs, however, have been more effective than IGOs in performing the task of technical assistance for development and welfare aims.

The conclusions of the present study tend to support these possibilities; however, the evidence in this case seems not to be as straightforward as one would expect. In addition, these thoughts are upheld in different ways by different organizations. For example, with its vast financial resources and influential policy papers on forestry, the Bank certainly has helped to generate greater interest in forestry matters, especially for rural development, simply because it has stated that it was interesting in funding that activity. FAO Forestry has helped develop interest and consensus through documenting the value of this new type of forestry through publications and discussions at meetings as well as through individual country projects organized by its Forests, Trees, and People Programme. CARE, on the other hand, has not directly developed consensus on the need for community forestry. But through its works it has provided important tangible proof of the value of community forestry, with IG0s pointing to CARE's (and other INGOs) work as illustrations of successful models to be pursued.

CARE, the only INGO in the study, has generally been quite effective in the technical implementation of its community and rural development forestry projects. The issue of determining whether CARE's overall impact was greater than the IGOs is far more problematic, however. CARE' impact has depended largely on the character and nature of the task activity. For example, CARE has been geared to promoting development for local communities. The planting of trees has also been a visible activity. Either the trees have been in the ground and growing or they have not. FAO Forestry, by comparison, has spent considerable time, not so much with local communities, but with government forestry agency officials and their staffs. Here FAO's activities of providing information or suggesting changes in forestry policy or helping to restructure the curriculum of a technical course *have not been as visible* as directly planting trees, but have been as essential. Of course, FAO Forestry has implemented field projects but has more frequently served as a technical support for the government, which, again, has been its principal target population. Actual technical achievements and the improvement of local communities has, then, become secondary to the training of government personnel, or at least providing the agency with new equipment. For example, it would not have been inconceivable for an FAO community forestry project to fail miserably in attempting to help the poor but still be viewed as a successful learning experience for forestry officials, especially if several new trucks were also acquired. FAO Forestry has been more of an organization of *suggested* than of *direct action.*

The Bank too has had to work through government agencies as its principal target population. The effectiveness of the Bank's projects has depended in part on the ability of the government to implement them, on how the project has been framed, and on how well it has been designed.

Without question, CARE's ability to ignore any serious political considerations has freed it to work on the technical objectives of its programs—but again, important political work has needed to be done that has been left to the IGOs. Helping the poor meet their forestry needs cannot be sustained without governmental support and capabilities, which the Bank and particularly FAO Forestry has tended to address. This issue of linked performances is discussed in the policy recommendations section below, and more conceptually in Chapter 6 when I discuss the role of interorganizational relationships and organizational fields.

Causal Relationships between Organizational Elements and Outcomes

In this section, I take the discussion a step further by outlining the possible cause-effect relationships that may exist between the organizational elements (technology, structure, and task environment and the observed organizational behaviors (commitment, appropriateness, and performance).

Table 5.3 presents summary explanations of the causal relationships between the study's causal elements and behavioral outcomes. The ensuing discussion is brief since it reworks many of the points established earlier.

Table 5.3
Causal Relationships for Organizational Behaviors

	World Bank
Commitment:	Bank is limited by investment-oriented technology and by the task environment desire of finance ministries for export revenues. Except in certain circumstances, Bank can't move its money through community forestry because demand is low and absorbative capacities weak.
Appropriateness:	Bank is limited by technology and task environment to only those countries willing and able to make social investments. It tends to exclude poorest of the poor countries.
Performance:	Bank is limited by technology's need for top-down blueprint project design. There is an inappropriate match between technology, structure, and task. Output task environment agencies are often incapable or unwilling to implement community forestry projects.
	FAO Forestry
Commitment:	FAO is dependent upon input task environment for political and financial support for community forestry. It is dependent upon output task environment for political support and demand for projects. Its commitment is relatively low since multilateral donors continue to support traditional forestry.
Appropriateness:	FAO is most dependent upon input task environments. It has a moderate level of appropriateness as a result of donor members' targeting needy countries while major funding agency is not.
Performance:	FAO is highly variable. Its performance is dependent upon uncertain interaction of technology, structure, and task environment. Its performance is negotiated between decentralized FAO unit, funding agency, or donor, and recipient agency.
	CARE USA
Commitment:	CARE's relatively strong commitment is the result of a good match between technology, structure, and input task environment. CARE is only capable of supporting community forestry. Funding is closely linked to CARE's strong performance and the desire of individuals, funding agencies, and donor governments to fund this activity.
Appropriateness:	CARE's high level of appropriateness is the result of its flexible, bottom-up technology, and CARE's decentralized structure is designed to work in poor countries. These characteristics are well suited to those countries that need community forestry work the most and that allow external agencies to control their own activities.
Performance:	CARE's high level comes from positive interaction between technology, structure, and task environment in achieving technical objectives. Its flexible, bottom-up, community-level technology and decentralized structure is well matched to community forestry tasks. It maintains complete control over the task, minimizing distractions from recipient governments and funding agencies.

World Bank

The interaction of the Bank's core technology and task environment has constrained its *commitment* to rural development forestry (except in India, where unique conditions have existed), though it has worked quite well in moving money, relatively speaking, when compared with the other two organizations. Earlier, we noted that community forestry as part of the Bank's total lending was very small. And without several very large loans to India and one to Brazil, the Bank's commitment would have been considerably smaller. In the most direct terms, because of the Bank's organizational arrangements, it could not move money efficiently through forestry, especially community and rural development forestry. The tasks of people forestry have not been well suited to the Bank's character.

There are several reasons for this. First, the Bank's core technology frames its projects in investment terms. Therefore, overall demand for rural development forestry projects has been low. Second, as a consequence, most governments have not been interested in forestry projects as social investments simply because revenue-generating ones make more sense politically and economically. Export-oriented projects or those that improve infrastructure are more popular among government and Bank officials. Finally, few countries (except for India and Brazil) have had the forestry-related infrastructure to absorb Bank loans of much size, making forestry a poor means of moving money.

This discussion leads into the causal factors for explaining the very low *appropriateness* of the Bank's work in community forestry. Only those countries with the political willingness and financial resources to make social investments are likely to have demanded these forestry projects. That is, only the wealthier countries in the Bank's task environment have been in a position to ask for substantial loans for this purpose. Still, a country has had to be wealthy, but not too wealthy. India's relatively vibrant economy, based on absolute GNP, and democratic ideals have allowed India politically to make social investments. Its large population size, however, has lowered its per capita GNP sufficiently for it to qualify for IDA's concessional loans. It is unlikely that India or just about any country would have been willing to invest in rural development forestry at the IBRD rates. Brazil has one such project. As a result of the conditions and very limited resources of IDA, it is difficult for the Bank to work where community forestry projects have been needed the most.

Regarding *performance,* once again community and rural development forestry projects sponsored by the Bank have been severely limited by its organizational elements. Since the recipient government forestry agencies have been responsible for project implementation, they must claim at least a portion of the blame for failures as well as credit for successes. As noted earlier, many forestry agencies have been weak and ineffective, hindering project successes. Those agencies' structures and professional training have tended to be inappropriate for implementing community forestry projects especially and have been in need of serious bureaucratic reorientation. Occasionally, structures have been created to simply absorb the money without any particular regard for other impacts.[2] Nevertheless, the control over project implementation has obviously affected the overall performance of Bank-sponsored loans, though the Bank would deny responsibility for this aspect. The Bank has *not* considered project implementation as part of its domain. Consequently, the project loan agreements have generally been more vital to the Bank than the actual impacts of projects.

Be that as it may, the Bank's project staff, however, has often been responsible for the project's design as well as its appraisal. The investment technology of the Bank is such that the staff members have to frame the project as a loan, which in turn leads the staff to devise a detailed, top-down, "blueprint" plan of action for the implementors to follow. To calculate economic and financial matters, a number of assumptions have been made regarding each project. Such assumptions have included types of tree species used, location of trees, number of trees planted and/or distributed, growth rates, yields, expected uses and prices, various costs, and so forth. To assure that the calculations and, consequently, the analyses remain fixed, the project specifications have often been quite rigid, certainly more so than for the other two organizations. This, as we have already noted, was perhaps the least advantageous methodology for engendering successful community forestry (or any other social welfare projects). The inflexibility, exacerbated by frequent inaccuracies in the project designs, has in part been engendered by the Bank's centralization. More accurate information could have been collected, and perhaps even greater flexibility facilitated, if Bank project personnel had been located near the actual site. Still, even this change belies the difficulty in adjusting the criteria necessary for framing the project in investment terms. In addition, since most of the projects have been constructed by project economists, economic investment criteria have tended to be more appreciated and more accepted as central to the Bank's operations than technical, environmental, and social information has, although there have been recent efforts to correct this problem. Still, the emphasis at the Bank has remained on moving money.

FAO Forestry

Unlike the Bank, FAO Forestry's behavior and performance in its forestry work has been more a function of task environment constraints than technology and structure. The characteristics of its technology have been overshadowed by the scope of its dependency upon others. FAO Forestry has used its technology—professional knowledge—to influence the course of events, but its approach has been more subtle, engineered through diplomatic measures and pressures. For all three specific behavioral outcomes (commitment, appropriateness, and performance), its efforts have been shaped mostly by negotiated events within its task environment, particularly between the donor and recipient countries.

Its *commitment* has been thoroughly shaped by the political desires and financial support of donor countries and funding agencies, as well as by the recipient countries demands for this technical service. Although FAO Forestry has played a role in shaping the political demand for community and rural development forestry through publications, seminars, conferences, and the like, it has remained dependent upon its task environment for its ultimate level of commitment. The principal reason that FAO has not done more in community forestry, especially in its Field Programme work, is that its major funding source for field projects, UNDP (United Nations Development Programme), 1980–90, has been concerned mostly with traditional forestry. UNDP funding priorities can be traced back to the recipient countries and the institutionalization of past projects, creating the desire for future projects of a similar nature. Consequently, new projects have tended to feed established structures created by old projects. Political relationships have been ensconced around existing traditional forestry projects, which have simply acquired more resources.

Most of the Field Programme's community forestry projects, however, have been underwritten by donor member trust funds. Even the tremendous rise in trust fund support for forestry that came in the late 1980s has not directly translated into proportionally more people forestry with the channeling of increased resources into forestry sector development more generally, including possibly some community forestry, as a part of TFAP funding efforts. Its Regular Programme commitment to rural development forestry, however, has been positively influenced by the financial and political support of SIDA (Swedish International Development Agency) and, more recently, other European governments.

The same general arguments help to explain FAO's moderate level of program *appropriateness*. Here, FAO has been more dependent upon support from individual donor countries than on multilateral funding sources. In addition, through donor countries as well as UNDP, community forestry projects have been packaged as a grant rather than a loan. Recipient governments have been more likely to accept a donor's offer to give money for community forestry projects than they would have been if the offer had to be repaid as an investment. This too has helped to explain the relatively greater appropriateness of FAO Forestry's program over the Bank's. A number of donor countries, especially from Scandinavia, have targeted community forestry projects where they were needed the most. But donor country trust funds during the study period constituted about 37 percent of FAO Forestry's Field Programme, and not all of that was allocated to community forestry (Brechin 1994).

As an international service agency, FAO Forestry must work wherever it is asked to. Thus, it has not been able to concentrate resources of its own on any particular set of countries or projects. UNDP must follow a similar protocol and distribute resources in a relatively equitable manner. This explains why FAO Forestry's work has been more evenly distributed than the Bank's or CARE's.

The highly variable *performance* of FAO-sponsored community forestry projects has been the product of the dynamic and unpredictable interaction among its technology, structure, and powerful task environment. Here again, project success has been contingent on the success of the negotiated agreements among funding agency, recipient countries, and the FAO professionals. It would appear that if the forestry agency is seriously committed to community forestry ideals, and the funding agency provides the right resources in a timely fashion, and the decentralized FAO professionals work well with the recipient and funding government officials, success will be enhanced. If the negotiated agreement or interactions falter anywhere along the line, disappointing results are likely.

Overall, in spite of its limitations, FAO Forestry has tended to have a more supportive technology, decentralized structure, and supportive task environment for its community and rural development forestry program than the Bank has had for its own. Consequently, compared to the Bank, FAO Forestry has been able to achieve greater overall performance for its program, especially when working with forestry agencies and their personnel.

CARE USA

Of the three organizations, CARE has been the most *committed* to community and rural development forestry. This is the result of CARE's character and its mission for humanitarian causes. Because of its decentralized community-development decision-making core technology and its decentralized structure, CARE has been much more likely to sponsor appropriate, flexible community-based projects. Community and other rural development forestry activities fit perfectly with these types of organizational elements.

Unlike the Bank, and especially FAO Forestry, CARE has not promoted other types of forestry activities. CARE has focused all of its forestry attention on community and rural development forestry. And while community forestry may appear to Third World governments as an unattractive investment, it is an ideal self-help, community level, humanitarian activity, which has also been of great interest to CARE's financial supporters. Here it becomes quite clear how the characteristics of the task itself influences organizational outcomes.

CARE's promotion of its community forestry work has been quite successful and was the organization's centerpiece development activity of the 1980s. With increasing forestry needs and the growing awareness of the problems of deforestation, private individuals and donor organizations (e.g., USAID) have allowed CARE to increase its commitment to community forestry by providing financial support for its efforts without many strings attached. The strong commitment to CARE has also been encouraged by its high level of performance. Funders obviously like to think their money is making a difference. Here, we have an organizational outcome, namely, performance, influencing another organizational outcome, namely, program commitment.

The high level of *appropriateness* of CARE's community forestry program has been the result of the positive interactions of its technology, structure, and task environment with the task. As indicated earlier, CARE's supporters have not been particularly willing to part with their money for those who are relatively well-off. This has forced CARE to work in the most impoverished countries.

The recipient governments have also played a role in CARE's program appropriateness. Although recipient governments could have said no to community forestry projects in their countries, there has been little reason to do so, since CARE's projects, like FAO's, have been financed by grants, not loans. In addition, governmental contributions to these projects have tended to be minimal.

The appropriateness of CARE's community forestry work has also been encouraged by its decentralized structure and decision-making technology, but in a "backhanded" manner. Because CARE's approach has demanded considerable control over its projects, only the most desperate countries have been willing to put up such a challenge to their sovereignty and competence. Clearly, CARE has worked in a type of task environment different from that of either the, Bank or FAO.

Finally, CARE's comparatively high marks for technical *performance* have come from the interaction between its flexible, "make it work," community-focus technology and from its semipermanent structures, the decentralized missions. CARE's technology has matched up well with the complex task of designing and implementing community forestry projects. Successful community forestry projects, as we noted earlier, must be geared toward the needs of local people and must have a certain degree of support for and involvement in the effort. These conditions are extremely dynamic, and the pertinent issues have varied from project to project and location to location. Consequently, there has been no single formula to apply. Community forestry is a type of project that must be constructed from the ground up and then frequently adjusted as needs arise. Although other PVOs may have been much more participatory and less heavy-handed in their development approach (see Van Wicklin 1990), CARE has certainly been more sensitive to this approach than the Bank or FAO has. More important, CARE has insisted on substantial control over project resources. This has allowed it, among other things, to quickly adjust projects to maintain the proper fit between the activities and tasks to be performed, more or less regardless of the government's position.

Like FAO Forestry, CARE, with its decentralized structure and centralized technical support from its home office, has helped assure flexibility in the project design and implementation with assurances of technical soundness and financial support. Unlike FAO Forestry, however, CARE's decentralized structure is semipermanent and not simply based on individual projects. Its structure is program based, not project based. CARE missions, once established, are in operation for a long time.[3] The semipermanent nature of CARE missions has provided greater program continuity and project support.

Because of its control over projects and its own funding sources, CARE has had the ability to continue or stop projects as warranted. It has been less constrained by preestablished project timetables or shared management obligations. CARE has terminated projects ahead of schedule when they were considered beyond salvation, for example, in Egypt in 1984. Through working out new funding arrangements by using its own resources and those of others, it has been able to continue and expand successful projects for years, for example, with the Majjia Valley windbreaks that lasted 17 years, from 1976 to 1992. (Its other programs, such as school feeding, have continued for decades.) In addition, through its decentralized program structures, flexible technology, supportive resource (input) task environment, and less-constraining project (output) task environment, CARE has continually experimented with new project ideas. CARE has had failures. But they have usually been quickly dismantled and the resources diverted to new ones. Overall, this approach has allowed CARE to maintain superb and innovative programs. Few international organizations have had this degree of control over resources, flexible administrative capabilities, and political autonomy.

Finally, as in each of the three organizations, CARE's project or output task environment has helped or hindered its performance in community forestry. Recipient-country governments, whether through direct action (e.g., elite monopolization of resources) or indirectly through established institutional arrangements (policies, laws, and bureaucratic procedures) have created the atmosphere for the basic level of support or lack thereof for community forestry. The government has been essential in providing fertile ground or barren rock for this new type of forestry.

CARE can search for and cultivate fertile ground more easily than most other organizational types. It can pick and choose its own projects and sites. This has allowed it, even while operating in countries unsupportive of forestry (e.g., Niger), to locate oases of support (e.g., competent individual forestry officials who were respected by the local people) to develop their projects. Also, at times, CARE has become a local community advocate and has run interference for communities against the government on important but controversial issues (such as distribution of project resources) in direct and pointed ways unavailable to FAO Forestry, or of little concern to the World Bank. Ultimately, if the situation has become completely untenable, CARE has simply stopped the program and even closed its missions. In short, CARE has had more financial and political flexibility in overcoming unresponsive recipient governments than the other two organizations have. This has allowed CARE to emphasize on-the-ground performance much more easily than either the Bank or FAO has.

Policy Prescriptions and Other Musings

Different organizations are different. This is hardly a revelation, but it is valuable to recognize that organizations have different strengths and constraints and, consequently, may or may not match up well with a particular task, which has its own unique set of characteristics. Should we worry much about this? I contend that we should, for the sake of improving success or minimizing inefficiencies in the development process. Moreover, there are policy prescriptions one can make, at least broadly speaking.

First, it would be foolish to lay out here a detailed set of recommendations after having taken such a broad, sociological view of these organizations and their community forestry programs. Consequently, the policy recommendations are themselves quite ecumenical. I offer them in hope of stimulating further discussion. At best, our goal here can only be to move closer to what might be sound courses of organizational action.

From this analysis it is clear that organizations are constrained structures; that is, they each are able to pursue certain types of activities more effectively than others. Organizations should be encouraged to identify their strengths and weaknesses and then take on those tasks most suited to them, much as we have already done with the three organizations discussed here. Next, I will offer two sets of recommendations that I have labeled first and second order—with the former being more fundamental and the latter more marginal in effecting change.

First-Order Recommendations

If we look a bit more systematically at the strengths and weaknesses of the three organizations regarding their work in community forestry, it appears that the organizational strengths of one, which also represents a class of similar organizations, or an organizational field (development banks, e.g.), match up well with the weaknesses found in another organization and field, (e.g., PVOs), and so on. What emerges out of this is essentially an argument for a *niche arrangement,* or greater specialization within development activities. Whether through generating common knowledge, establishing more competitive forces, or mandating interorganizational collaboration by some superagency, the idea is to build upon the strengths of particular organizations and have them work on the development package or piece they do best. In addition by raising the issue of organizational fields, discussed more in Chapter 6, one can begin to speculate more thoughtfully on generalizing the findings from these three organizations to many others and from community forestry to other similar tasks. Table 5.4 outlines the particular strengths of each organizational type and the way in which each might specialize in certain activities around community forestry, or any other socially dynamic and unpredictable task. The concept of cooperation and coordination in development assistance is hardly new; however, most of this work has focused on cooperation within and among governments (see, e.g., Berg and Gordon 1989) and less directly among the development organizations themselves. There are a few exceptions, however. In their review of eleven multilateral development agencies, the Danish aid agency DANIDA came to a similar conclusion, arguing for greater specialization and coordination along organizational strengths (DANIDA 1991; see also Fairman and Ross 1994).

In pursuing a particular task like community forestry, organizational niche arrangements might simulate more effective development efforts. By way of example, the, World Bank or other multilateral lenders might want to focus their efforts on macroeconomic issues to provide a more supportive economic and political environment for encouraging community forestry activities. Another possibility could be in the area of sectoral loans to finance the redevelopment of a nation's forestry program that might better encourage community and rural development forestry activities, as long as there could be assurances that greater social and environmental damages would not result from the need to repay those loans (see Korten 1994). Development banks might even provide loans to a specific community forestry project if the project has already been successfully established and is constrained only by lack of capital.[4]

FAO or other technical assistance organizations should concentrate on what they do best. In community and rural development forestry projects, FAO ought to focus on assisting governments and their forestry agencies with creating the necessary bureaucratic elements to successfully manage people forestry. FAO might be asked to suggest appropriate forest policies or to train forestry personnel in the art and science of community forestry. Under the right conditions of supportive donors and recipient agencies, FAO might be a very appropriate choice to design as well as implement community and rural development forestry efforts on the ground.

Finally, CARE USA, or other PVOs, may best serve the task by both designing appropriate and then actually implementing community forestry projects for local communities. In particular, many of these organizations excel at implementing new and innovative projects or approaches to development. It is quite possible that some of their efforts might become mod-

Table 5.4
Organizational Niche Argument:
Matching Organizational Field Strengths to Community Forestry
or Similarly Dynamic Tasks

World Bank and other development banks:
- Macroeconomic policies (e.g., structural adjustments)
- Sector loans and policy statements
- Financing of established but capitally constrained projects

FAO Forestry Department and other technical assistance agencies:
- Professional agenda setting, legitimacy, and diplomacy
- Technical assistance
- Bureaucratic reorganization and educational and training programs
- Implementation of forestry-sector development made necessary by development banks

CARE USA and other PVOs:
- Designing and implementing complex (e.g., socially dynamic) tasks for local communities
- Innovative project models/approaches for agency emulation/adoption
- Direct and immediate development assistance for poor communities

els worthy of emulation by forestry agencies themselves. In addition, PVOs tend to be at their best when working directly with local communities for the people's benefit. PVOs often help to ease the immediate human suffering while other development activities create the necessary social and physical infrastructure to improve the general welfare for the future. Thus, we see that temporal niches might emerge as well. Ideally, the overall goal would be for the government to become competent enough in community forestry matters so that it could take over projects and approaches that were pioneered by the PVOs. This would allow these types of organizations to continue their cutting-edge perspective and innovative approach to development while the existing efforts become institutionalized.

Second-Order Recommendations

Second-order recommendations represent additional ways in which one might encourage successful organizational efforts in community and rural development forestry, but at a more modest level. For example, the flexibility and decentralized decision-making processes of CARE and other PVOs should be enhanced with funding that has as few strings as possible. USAID has been promoting this approach with considerable success. Block grants give PVOs the leeway to utilize their strengths. Accountability can be maintained through various monitoring mechanisms and by tying future grants to past performance. Here attention is paid to final outcome (or periodic markers, e.g., yearly reviews) as opposed to detailed oversight of every decision and dollar disbursed.

One of the most serious problems facing CARE and other PVOs is their lack of significant capital. Obviously, it would be ideal to link PVOs with the development banks, which have lots of capital but possess very limited capacity for socially dynamic tasks. Unfortunately, the very different nature of these organizations, particularly the banks' investment approach and dealings directly with national governments, greatly complicates forcing such a working arrangement. CARE tried working directly with the World Bank in the late 1980s on forestry in Uganda, but quite unsuccessfully. Others have explored linking the Bank with PVOs (e.g., Fisher 1993; Nelson 1991; Paul and Israel 1991), but the signals so far are, at best, mixed. It appears PVOs may be best served with support from their own governments and private donors, although I still hold out some hope of a bank linkage if the proper conditions can be identified and established. We need to find better ways of making use of the enormous capital controlled by the banks, allowing them to better support the implementation magic of PVOs.

Almost all of the Bank's loans to community and rural development forestry were from IDA, the Bank's soft-loan, low-interest-rate affiliate. This suggests that governments are generally uninterested in investing in community and rural development forestry, unless the opportunity costs are quite low. Still, the necessity of repaying the loans keeps many of the countries with the greatest need from investing in these types of activities in the first place. This is very unfortunate. It would be ideal if the available capital could match the demonstrated need, since those countries that can afford and manage Bank loans tend not to be those in the most desperate straits.

A partial response would be, on the one hand, to increase the IDA pool through higher levels of replenishment by donor members, in part for the explicit purpose of social investments like community forestry and, on the other, to assure that capital at the best terms is always available for these kinds of projects in the countries with demonstrated need. Perhaps such projects could have a negative interest rate, where part of the loan is written off over time. Pushing this notion even further, the amount of the negative interest rate could be tied to the success of the project, with greater success linked with a greater write-off. This might provide incentive to a country to assure the success of the investment. With financial savings at stake associated with the project, and the usual lack of institutional capacity, perhaps these impoverished governments should be encouraged to invite technical assistance organizations like FAO, or PVOs like CARE, to assist in the design and implementation of the project. In another variation, perhaps the Bank would offer the government a lower (or even negative) interest rate if appropriate PVOs or technical assistance organizations were involved in the design or implementation of the project, which the Bank would simply finance.

Unfortunately today, discussion of FAO necessities a discussion of TFAP.[5] Although TFAP has become a major project of FAO Forestry since the late 1980s, FAO Forestry is larger than TFAP. There are many complex reasons and perceptions behind TFAP's difficulties. The original critics of TFAP said that it was nothing more than a scam to increase tropical logging; others suggested that problems emerged when TFAP failed to gain the full attention of chief political leadership necessary to instill the wholesale change and coordination among different units; and some felt TFAP did exactly what it was suppose to do, namely, bring more resources to forestry management agencies (see Brechin forthcoming; Lohmann and Colchester 1990; Winterbottom 1990). The following recommendations should not be seen as in response to, nor in ignorance of, TFAP, but rather suggestions FAO may want to pursue regardless.

The recent (1988–90) rise in donor trust fund support for mixed projects found in my data reflects the financial commitment to the TFAP process. It is unclear how much of these resources was actually used to promote community and rural development forestry. Since trust fund support has historically made up the greatest portion of FAO funding for community and rural development forestry (64% from 1980–90) (Brechin 1994), it would be unfortunate to have that commitment blunted rather than enhanced by the TFAP process. I hope that donor nations will maintain or even increase their commitment to people forestry.

During the 1980s, at least, UNDP's commitment to this innovative forestry approach was strikingly low (32%). Much more support needs to be shown by UNDP, given the size of its forestry funding in general, such support would translate into a major financial boost for rural development forestry projects.

Finally, FAO Forestry's greatest success in community and rural development forestry has been its FTPP. Under skillful leadership and smart financial support, this FAO program has done much to stimulate interest among and to train government officials in community forestry. In addition, FTPP has been able to channel benefits directly to local people and communities through constructive requirements placed upon participating government agencies. This program should be handsomely supported and reshaped as necessary to keep it vital.

Implementing a Niche Arrangement

The second-order recommendations just outlined could be implemented fairly straightforwardly. Here political will is likely more a barrier than know-how. The first-order recommendations on greater specialization, however, are a different matter. How do we get multiple organizations to coordinate their skills and resources for a given task? This is, for the most part, uncharted territory. To some, this discussion of coordination may sound a bit like the reinvention of TFAP. Yes, TFAP has had its share of problems, but the concept of coordinating specialized development assistance remains reasonably sound. For our discussion here I offer two sociological possibilities for encouraging this: voluntary compliance and coerced coordination.

Promoting voluntary compliance among development organizations calls for a change in society's expectations of them. This would likely be accomplished most effectively by changing the institutional organizational environments (see Chapter 6) in which many of these organizations find themselves. From a political science perspective, change could come as the result of expert agreement among members of epistemic (or scholarly) communities (Haas 1989) that might form around the utility of a niche-arrangement approach to development. Similarly, with support from scholars eventually even a type of regime (or agreement on proper behavior) among member nations might emerge, organized around the necessity of the approach and how best to proceed. Hall (1993), however, argues convincingly that a shift in policy approach of this magnitude would take more than just experts, but politicians as well. Such a coalition to push for a shift in approach would be unlikely until the present arrangements repeatedly come up short.

Within sociology this approach represents more of a cultural perspective, wherein conventional wisdom or worldviews, specific practices, routines, and so forth become accepted or institutionalized within a culture so as to become a regular part of life. Here, with time, agreement, and continued discussion, it would be only common sense to have a niche arrangement. In short, the results of research, writing, and discussion on the strengths and limits of development agencies, given the types of development tasks to be performed, would all need to become codified and then accepted in some manner. Once this happened, discussion among the agencies themselves could likely ensue, and eventually an implementation of the niche arrangements would form in response to pressures from each organizational task environment as well as institutional environment. "Voluntary" compliance would then emerge.

Under the general rubric of voluntary compliance, greater specialization could also be encouraged through a more competitive bidding system. Here, development organizations might compete for business with "bids" going to the organization or organizations that provide the right service at the right price. Certainly it would be far removed from a pure-market situation, but rather would be an attempt to find ways to incorporate marketlike incentives into the institutional world of development assistance. How to organize such an arrangement would be a challenge, given how highly diffuse and uncooperative most national development programs are. Obviously, such an approach requires a greater role for international organizations.

Information too would have to be available so that informed choices could be made: on the work required and the skills and performance record of the bidding organizations. Issues of organizational management and of sovereignty would need to be dealt with. Unlike most other development organizations, NGOs are much better able to respond competitively, because of their numbers and relative substitutability. These organizations tend to be more efficient and effective since they must maintain closer connections between their outputs and inputs in order to survive. Of course, that this competitive approach assumes customers and service providers are all looking for efficient outputs I think is questionable. Still, finding ways to increase competitive forces within the largely institutional environment of most development organizations should be investigated.

Instead of voluntary compliance, another way to coordinate development assistance among agencies would be through some kind of coercion. In an international setting, we would be wise to take our lead from Garrett Hardin and mutually agree to coerce ourselves. This could be achieved essentially by the same process just described above. But instead of there being voluntarily collaboration among agencies, forces within the broader institutional environment would require the organizations to coordinate their activities. A possibility here would be to establish a new international organization that would oversee the coordination of activities. This organization might arrange development "deals" to match development needs of a specific country and finances from a donor agency and implement plans by some agency or organization, all coordinated around the particular task to be pursued. The agency would work within a negotiated framework, but the matching agency would have veto power for arrangements it did not view as appropriate and would have a way of punishing parties that did not negotiate faithfully.

Such a coordinating agency is not entirely far-fetched. A prototype may be the Multilateral Fund for the Montreal Protocol. This organization was created recently to assist in the transfer of development assistance from the Northern to the Southern Hemisphere as part of the Montreal Protocol on ozone. In compensation for the loss of CFCs to industrial processes, developing countries are to be compensated through the transfer of funds from developed countries (DeSombre and Kaufman 1994). The Multilateral Fund for Montreal Protocol, especially through its executive committee, has been attempting to match parties and projects. It has also 'vetoed' potential transfers of funds that it thought were not true to its mission. Obviously, this type of arrangement would need to be explored further, but it remains an intriguing possibility.

Chapter 6

A Theoretical Review: Toward a Sociology of International Organizations

This study has investigated the effects of three organizational elements—technology, structure, and environment—on the behavior and performance of three international organizations and their work in community forestry. Organizational *technology* refers to the skills possessed and the procedures used by an organization as it works to achieve an end. Organizational *structure* reflects how the organization's work (or its division of labor) is formally distributed within the collective and how it is controlled. Organizational *environment* represents the forces external to the organization itself; it implies that no organization is self-sufficient. For this study, several environments were identified. For example, three nested levels of organizational influence were relevant: micro-, meso-, and macrolevels. The macrolevel embraces the broader economic, social, cultural, and political forces within society and includes the general distinctions between technical and institutional environments. The mesolevel reflects the direct organizational relationships surrounding particular organizational activities, that is, the task environment. And the microlevel environment is the particular, detailed setting of the organization's outputs, such as particular development projects.

The Organizational Environment

The environment in organizational theory represents all the forces and elements that are "outside" the "boundaries" of the organization. It includes, as Perrow (1986) bluntly says, "anything out there" (192). Attention to the environment of organizations is a rather new phenomenon in the organizational literature. Rather than view organizations as self-contained systems, separate and distinct from the larger environment, theorists began to more consciously view them as incomplete systems and as "loosely coupled" systems (Ashby 1968; Buckley 1967; Glassman 1973), less distinct from and more dependent upon elements found within the larger environment. As a result, organizations are forced to interact with their environments to obtain the resources and opportunities they need to function. From this perspective, organizational environments have become viewed somewhat more seriously, with concrete implications for organizations (Aldrich 1979; Katz and Kahn 1966; Lawrence and Lorsch 1967; Meyer and Scott 1983; Pfeffer and Salancik 1978; Powell and DiMaggio 1991; Thompson 1967). In particular, a couple of important theories emerged from this resource view of organizations: resource dependency theory (Pfeffer and Salancik 1978) and institutional theory (Meyer and Scott 1983; Powell and DiMaggio 1991).

This is not to say that earlier students of organizations did not acknowledge the existence of the organizational environment or its importance. As noted by Perrow (1986): "It was always there in organizational theory, from Weber on" (178). This is clearly evident in Selznick's work (1949) as well as those already mentioned, and in Lawrence and Lorsch (1967), Thompson (1967), and a host of others (Perrow 1986, 178). What was lacking in these earlier works, however, were more distinct conceptualizations of how to think about the environment. Significant intellectual attention by theorists is now focused on more precisely understanding the environment and how it relates to organizations.

Scott (1992) presents three basic levels, essentially discipline based, on which one can approach the subject of organizational environment: social psychological, structural, and ecological (126). Each are valid and collectively help to explain the large variation in the conception of organizational environment. In this study, the structural and ecological levels have been most relevant. We have made use of the structural level when discussing contingency theory, while the ecological level has served as the foundation for discussions of interorganizational relationships, such as organizational sets, fields, and communities.

The three physical levels of environments I have posited—micro-, meso-, and macrolevels—are scale related. The three organizations essentially functioned within several environments, each affecting its behavior and performance of a given task (see table 6.1 below).[1]

Table 6.1
The Study's Level of Environments

Macrolevel: Societal influences (e.g., broad technical and institutional)
Mesolevel: Task environment (e.g., intimate organizational interactions)
Microlevel: Task's (output's) environment (e.g., local settings of projects)

Table 6.1 suggests a number of conclusions about the level of environment and its consequences for organizational theory. Some of the relevant observations from the details of our case studies are presented below. Before attending to them, however, it might be instructive to consider how these different levels may interact with and influence organizations more generally. For example, these three levels are quite similar in concept to Hall's (1993) notion of how shifts occur in state policies. Hall too presents three levels of policy actions: the microlevel settings of policy implementation, the mesolevel changes that affect the policy instrument itself, and the more macrolevel general goals that support particular policy approaches. Change in policy approaches, or "policy paradigms," argues Hall, come when mounting empirical evidence on the ineffectiveness of standard approaches punches enough holes, not only in the prevailing policy instruments but in the overarching goals as well. In response to the inconsistencies and failures, newly formed coalitions emerge to promote new goals and approaches.[2]

With Hall's example, one has the accumulation of microlevel failures or inconsistencies that either can't be tolerated or explained and that eventually bring down the whole deck of cards. Consequently, then, local-level events are important in the eventual outcomes of organizations, though the pieces of the puzzle may take decades to accumulate. The current political troubles of the World Bank may indeed be partially the result of its not paying sufficient attention to the microlevel results of its lending practices, such as environmental impacts. Finally, although microlevel events will likely influence macrolevel developments, the reverse must also be expected. Here, changes at the macrolevel obviously can alter what happens at the meso- and microlevels. New information and change will go back and forth among the levels in a never-ending game of institutional ping-pong. Hence organizations need to pay close attention to the effects their activities have on the ground.

Microlevel Environments

The local environment of the particular community forestry task, or the microlevel environment where the organization's output is placed, has been key to this study. I have argued that although the organization's outputs have characteristics, such as the particulars of the community forestry package, so does the environment where this output is placed. The output's environment likely overlaps with the notion of contingency theory, that is, a stable or turbulent environment, but contingency theory has not dealt much with multiple levels of analysis (Pennings 1992) where the mesolevel task environment or macrolevel technical/institutional environment may also have been stable or turbulent. I am attempting to add greater specificity and integration to the analysis, because I am convinced that an organization's internal characteristics and its work in all three levels have an enormous influence on its activities, like community forestry. Certainly the particularities of the microlevel environment will be quite different for different types of outputs.

It is probably best to think of environmental conditions more as continuous rather than dichotomous variables. Regardless, each of the three organizations faced complex and highly unpredictable microlevel environmental conditions rather than stable, predictable ones. Each of the organizations performed differently in this kind of microenvironment. CARE performed very well with its relatively flexible outputs. FAO sometimes did as well, but it depended on a host of conditions largely determined by the specific interactions at this level and by forces within its task environment at the mesolevel. In general, the Bank's outputs performed poorly at this level because of the mismatch between the characteristics of its outputs and the needs of the microlevel setting. It has performed very well, however, at the meso- and macrolevels, which have been more relevant to its survival.

Mesolevel: Task Environment and Resource Dependency Theory

According to Dill (1958), an organization's task environment represents for the most part the other organizations with which that organization must interact to obtain its inputs and to distribute its outputs. I argue that task environment is to the organization the most directly interactive subset of elements found within the larger technical or institutional environments. Scott (1992, 133) places task environment only under the discussion of technical environments. I disagree with this; I view task environment as those other organizations most interactive with the focal organization's work within its relevant organizational set. Dill did not have institutional theory to guide his original description of task environment. To me, it is a type of environment based upon intimacy, or high degree of interaction, which emerges from the organization's work, regardless of its more technical or normative setting. Consequently, task environments can be found in both technical and institutional or any other environments that are based upon other general characteristics or properties. The concept of organizational fields (DiMaggio and Powell 1983, 1991) too can represent organizations that interact closely with one another. However, I restrict the notion to mean organizations that are similar to each other as an organizational type, such as PVOs, banks, and UN technical assistance organizations.

The open system perspective of organizations reflects their need to enter into exchange interrelationships with other organizations. Consequently, interdependencies develop. Inequities in exchanges, however, may create power-dependency relationships, as noted by Emerson (1962), among the members of the organizational set which can significantly influence organizational behavior and performance.[3] Similarly, Thompson (1967) considered task environments as a serious source of constraints to any organization (30).

Pfeffer and Salancik (1978) developed the notion of inequities of exchange relationships into a rich perspective known as *resource dependency.*[4] They stated that organizations enter into agreements cautiously and develop strategies and mechanisms to minimize their dependencies. Some of these methods are discussed below.

At a general level for individual organizations, the environment, whether technical or institutional, represents a set of greater or lesser uncertainties, risks, and opportunities with which organizations must cope. This is the case regardless of whether the perspective is focused on goal attainment (rational perspective) or organizational survival (natural systems perspective). Organizations and networks attempt to manage their environments, generally through conscious adaptations and the deployment of "offensive" and "defensive" strategies.

Organizational theorists over the years have identified a number of such strategies, known as "buffering" and "bridging" strategies, which organizations use in dealing with their environment (Scott 1992; Pfeffer and Salancik 1978; Thompson 1967). Most of the strategies identified to date have been for those organizations found in technical environments, particularly in manufacturing and other business organizations. We review briefly some of those strategies and a few others not widely discussed in the literature.

Buffering strategies were first outlined by Thompson (1967) as mechanisms that managers use to "seal off" the organization's technical core, or production system, from environmental disturbances. Such strategies consist of ways of protecting the technical core, usually by manipulating the organization's inputs or outputs. The basic buffering strategies listed in the literature are coding, stockpiling, leveling, forecasting, and growth. Several of these, such as coding and stockpiling, seem more applicable to manufacturing than to other types of organizations. Certainly, examples can be found that are relevant to this study. For example, CARE and FAO maintain, or "stockpile," a large roster of qualified applicants from which to select needed professionals. The roster is organized, or "coded," by the training, years, and place of experience (and similar qualifications). Although these strategies exist, they do not provide tremendous insights in understanding these organizations.

A buffering strategy with more relevance to this study is leveling. Leveling strategies attempt to reduce fluctuation among inputs and outputs, that is, to reduce the feast-or-famine problem. Leveling strategies used by the three organizations include finding a more reliable or diverse set of resource suppliers and attempting to stimulate demand for outputs. Although sales and incentives are obvious business examples, the use of publications, conferences, consultants, mission teams, task forces, and linked programming are examples of demand enhancement strategies used by the organizations found in this study.

If buffering strategies are classified as "defensive" strategies geared toward protecting the organization's technical core, *bridging strategies* are "offensive" ones, implemented to adapt to environmental conditions. These changes tend to be directed at organizational boundary maintenance rather than at the protection of an organization's technical core. Or stated another way, organizations attempt to protect their technical cores by eliminating the source of disturbance within their task environments.

Pfeffer and Salancik (1978) listed the basic bridging strategies: bargaining, contracting, cooperation, joint ventures, mergers, associations, and government connections. Thompson (1967) remarked that organizations with differing technologies tend to utilize different bridging strategies (42–44). For example, organizations with mediating technologies, like the Bank, attempt to adapt to the constraints found in their task environments by increasing the size and type of population served—that is, they grow. Similarly, organizations with intensive technologies, like CARE and FAO Forestry, tend to reduce disturbances by attempting to gain greater control over the object of work. Some of the evidence from our three case studies certainly support these behaviors, and they are points to which I will return.

Much of the preceding discussions on task environments and the organizations' ability to control vital input and output resources falls within the rubric of resource dependency theory. Considerable explanatory power can be found in the variation among the three organizations in (1) the degree of control held by their respective task environments and the consequences for organizational behavior and performance, and (2) the ability of each organization to manage their respective task environments through defensive and offensive behavior. For example, we noted that FAO particularly had difficulties in managing its task environments, while the Bank and especially CARE fared much better.

Distinguishing Input from Output Task Environments

In understanding this study's findings, I found it helpful to view task environments as quite varied and textured yet divided distinctly into two basic groups: input and output. In each case, the organization often faced a very different set of forces and problems, requiring different solutions. CARE generally had few serious problems with its output task environment, which it was able to more or less control through the careful selection of countries it worked in. If host governments were unwilling to play by its rules, CARE simply moved on. Its ability to control where it sets up shop has given CARE a type of administrative flexibility and power unparalleled by the other two organizations. In addition to location, CARE continued to maintain its performance by selecting only the programs, specific projects, and sites in which it felt it would be successful, that is, by selecting supportive meso- and microlevel environmental conditions for its operations and outputs. CARE also managed its output task environment by restricting its activities to only those most impoverished countries that would allow it to maintain near-complete control of its affairs. Host governments approved each project before implementation, but there were rarely serious holdups. CARE maintained additional clout through its larger operation in food distribution, which was part of what I call "linked programming." Food aid has been CARE's largest humanitarian activity. Once established in a country, CARE often attempts to build upon its food distribution activities by augmenting them with more development-based programming, such as agricultural improvements or community forestry. This linked programming is a way for CARE to build upon its existing operations in countries where its efforts are needed and the conditions are most favorable for its success. If necessary, CARE could use its influence as a major supplier of food resources as leverage to promote its development programs (as well as call upon the influence of the U.S. government if the project is cofinanced by USAID).

As I discuss in greater detail below, CARE's control over its performance has directly affected its relationships with members of its input task environment. Its inputs were closely linked to the performance of its outputs. By maintaining its reputation as an efficient and effective private relief and development organization, CARE has continued to raise money from individual donors as well as through grants from governments.

The Bank and FAO have attempted to manage their output task environments through stimulating demand for particular types of projects by using a variety of techniques. The Bank tended to push its programs by policy sector statements, various activities like country sector studies, and economic mission teams. These methods reflect the unique ability of the Bank to develop and carry out its own policy initiatives. They are also a way to stimulate new demand for its services by, in essence opening up new market areas. The Bank's initiative in forestry in 1978 can be seen in that light. In the mid-1980s, the Bank, with the cooperation of several other organizations, spearheaded TFAP. The basic purpose of TFAP was to encourage governments to invest more resources in the forestry sector as a means of achieving sustainable development. Thus, TFAP can be considered a kind of demand stimulus project as well. FAO Forestry also has tried to stimulate more demand for its services, but usually through the use of diplomacy, such as conferences, meetings, and publications.

The influence of the *input task environment* of the three organizations, as well as the success of their management strategies to control it, has also varied considerably. The Bank, for example, was rather insulated in the operations of its daily activities. But for its continued growth, the Bank was very dependent upon its donor members, particularly its largest shareholder, the United States. Although the Bank is self-supporting, and in fact profit making, it periodically asks for increases in its subscription rates to increase its level of financial service to recipient members. As a consequence, the Bank tries to remain in good standing with its donor members and is very sensitive to criticism by them. Here the Bank has managed donor task environments by attempting to directly address the areas of criticism through structural changes. Yet, the changes in structure have affected the Bank's technology or actual performance very little. The creation of the Bank's Office of the Environment in 1970 or its Evaluations unit in 1975 were made with virtually no impact on daily operations. During its 1987 reorganization, as the Bank attempted to make structural and procedural changes to respond to the increased need for structural adjustment lending, the Bank received enormous criticism from environmental and other groups about the negative social and environmental impacts of its lending programs. In an effort to appease members of the U.S. Congress, the administration, and numerous private environmental organizations, the Bank made a number of adjustments, including the expansion of the Bank's environment department (see Rich 1994). It remains to be seen how effective this reorganization will be on the Bank's operations and performance (Brechin, forthcoming). Although the country's ability to repay loans carries some weight in the Bank's and the international banking community's willingness to participate in the lending process, the benefit-cost analyses conducted by the Bank staff has had enormous influence on the viability of many socially relevant development projects. The belief in the infallibility of the benefit-cost tool has become quite internalized as part of the professional culture of project economists.

CARE has managed its input task environment through very effective advertising campaigns and promotions, humanitarian appeals, and, most of all, its generally efficient and effective performance on the ground. It does not take its self-promotions lightly. The public relations and fund-raising staff composed roughly half of CARE's total headquarters staff and budget during the 1980s. CARE has won professional awards for its promotional activities, including the Public Relations Society of America's 1984 Annual Silver Anvil Award for the "CARE for the Earth" campaign (CARE 1984).

CARE has also maintained its autonomy, while attracting financial support from government donors with cofinancing and matching-grant arrangements. These arrangements have been very attractive to donor agencies and has allowed CARE to extend its work. The task environment management strategies of the three organizations are presented in table 6.2.

Macrolevel: Technical and Institutional Environments

Of particular relevance to our discussion of the behavior and performance of the three organizations is the basic distinction between *technical* and *institutional* environments (see Meyer and Scott 1983; Scott 1992, 133).[5]

Briefly presented, Scott (1992), based upon his earlier work with Meyer,

Table 6.2
Input and Output Task Environment Management Strategies

	Input	*Output*
World Bank	Adjust structures Maintain professional quality of microeconomic analysis	Stimulate program and project demand
FAO Forestry	Adjust structures and technologies	Stimulate program and project demand
CARE USA	Maintain performance and promotions (including promotion of performance) Obtain matching grants	Limit activities geographically and by program type

states that organizations located in more technically oriented environments tend to be rewarded for effectively and efficiently producing their outputs (133). Clearly, this is geared mostly to marketplace organizations. As an ideal type, Scott (1987) notes that these organizations are those "dear to the hearts of the classical economists" where rational structures reign supreme (126). I have argued that certain nonprofit PVOs fit better within technical environments as well since their well-being hinges significantly on their actual performance rather than on their ability to manipulate normative symbols.

By contrast then, organizations in institutional environments are rewarded not so much for their efficient and effective production of outputs, but rather for their conforming to established procedures, rules and norms, that is, for having the correct structures and processes. While the desire for a favorable bottom line generally drives those organizations found in a technical environment, it is the desire for legitimacy and political support that tends to be the driving force for organizations within institutional environments. Organizations found in more institutional environments, such as mental hospitals, public schools, prisons, the military, churches, professional organization, and so forth, tend to be those whose outputs are less likely to be placed in competitive markets or difficult to measure. As we will see later in this study, many of the IGOs appear to be lodged in this type of environment. This is not to say that economic resources are not important to these organizations. Rather, these resources tend to be more con-

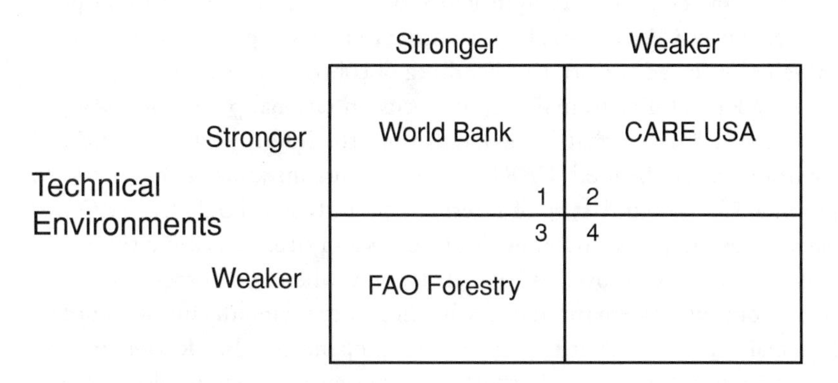

Fig 6.1. Technical and Institutional Environments: The World Bank, FAO Forestry, and CARE USA

nected to issues of legitimacy and political support and are less directly tied to level of output performance. In institutional environments, organizational survival is more loosely coupled with efficient production.

Instead of treating technical and institutional environments as simple dichotomous variables, Scott (1992) suggests placing each variable on a stronger-to-weaker continuum and to construct a fourfold table: "strong and strong," "strong and weak," "weak and strong," and "weak and weak." This interaction typology provides a useful framework for analyzing and comparing organizations. In particular, it provides considerable insight for understanding the behavior and performance of the study's three organizations (see fig. 6.1).

This typology provides a useful way to explain the variation found among the three organizations, especially when placed in a fourfold table. CARE provides the most straightforward illustration, so we begin with it.

CARE USA: Strong Technical Environment

Figure 6.1 shows CARE with a relatively strong technical, but relatively weak, institutional environment. Its strong technical environment is a result of CARE's need to compete with other development and relief organizations for scarce voluntary donations from the general public and from private organizations and foundations to support its overseas work.[6] Its success at obtaining funds from private citizens and organizations, and in part its ability to receive grants from donor governments, rests principally on its reputation as an efficient and effective community development and relief organization. Its inputs are tightly linked to the performance of its outputs. Consequently, CARE pays close attention to issues of performance overseas as well as to its administrative handling of contributions at home.

CARE's ability to maintain its focus on rational production is aided precisely by the fact that its demands from the international institutional environment are limited. CARE is a private organization. It has not been grounded in the same type of international milieu as IGOs like FAO and has not been required to follow those norms and rituals as explicitly. When it is invited to work overseas, it will go only where the host government, that is, output task environment, will allow it to maintain almost complete financial and administrative control over its operations. In addition, once in a country it will choose only those projects that it is convinced it can perform well and that will be supported and approved of by its *own* contributors, which are part of its task environment. These are luxuries afforded to neither the Bank, nor especially FAO.

The relative advantages that CARE possesses, however, are not without costs. Such so-called advantages limit both the geographic and programming focus of its work. Although it is unable do everything everywhere, there is obviously still plenty for CARE and similar organizations to do. But, as we already noted, the overall impact of CARE's development work can be questioned.

I would argue that the lack of the institutional environment's control over CARE's activities correlates with overall limiting impact. This suggests that organizations found mostly within institutional environments can have a significant impact on the institutional environment itself. Since CARE can largely ignore the elements of the international institutional environment, those elements, in turn, can largely ignore CARE. CARE itself cannot do much to change a government's opinion about community forestry, let alone the world's. It was not created for that role, and it has since avoided taking it on to maintain its overall technical performance, which then enhances its ability to grow and survive in its own more technical environment. At times, however, organizations of the more institutional environment have used CARE's (or other PVOs') success as proof of the viability of innovative thinking.

The World Bank: Strong Technical and Institutional Environments

The Bank, as shown in figure 6.1, exists in both a strong technical and strong institutional environment. As a consequence, it is the most complicated case to illustrate. One reason is that its dual environmental qualities immediately raises the question, performance for whom? Or, Which environmental elements should one address the most closely, and why? Obviously, the Bank has been doing something right. As an organization it has been very successful, especially since the 1960s.[7] We noted in Chapter 1 that the growth of the Bank over the last generation has been phenomenal.

The Bank has evolved into the chief development bank for the world community. Here the Bank addresses two critical cultural norms of the international institutional environment: the issue of economic development, or modernity, and the notion of a world community of individual sovereign nations. It is a "public" bank of the world community, dedicated to financing the development needs of sovereign nations. But the Bank was also created, in part, to tap into the resources found in the economic marketplace—that is, within a more technical environment.[8]

To perform well in its technical environment, the Bank must operate efficiently and effectively in the production of bankable project loans as outputs. In a blending of both technical and institutional conditions, the Bank has invested considerable time and energy in creating the image of foremost professionalism in the formulation of project loan packages. Only the world's most competent benefit-cost economists are hired to transform an ordinary development project into a bankable one. Within the Bank the creation of a sound appraisal is taken with utmost professional seriousness and is usually quite narrowly embedded in economic benefit-cost criteria. The point here is that the Bank has worked hard to create and protect its image as an extremely able and competent organization. This is especially important to the members of the Bank's task environment, primarily its borrowers and the international banking community. The Bank's image as a truly professionally competent organization has become quite institutionalized to a level of near-mythic proportions.

The Bank's competence has gone unquestioned among its most important collaborators while being severely criticized by complete outsiders. What has been more important to members of its task environment, however, has been its generation of loan agreements. The efficient creation of new agreements has been the true goal. Consequently, the Bank's professional staff is closely supervised, in part to maintain the quality of the loans, but more important, to push for agreements. As a result, the Bank's centralized, hierarchic structure of professionals appears quite rational. *Hence, the goal that has been maximized is the rationally efficient production of "bankable" loan agreements, not the actual impact of the loans when implemented.* The need for the Bank to find ways to move its money to continue its growth and to enhance its survival, explains its willingness to find creative solutions for financing a project—and just as important, it explains the Bank's adamant unwillingness to adjust those that are not working. The need to move money also explains the intensive efforts to develop new demand for its projects loans. Thus, we see how the Bank becomes most committed to those programs and countries that move its money most efficiently, not necessarily to those most necessary or needy.

More generally, because of the its strategic position in the international financial markets as a conduit between investment opportunities in LDCs and the world's major banks and investment companies, the Bank has garnered considerable support. The profits of many of the world's leading financial centers have depended, in part, upon the Bank's ability to open up investment opportunities in LDCs and to keep them financially solvent and economically viable.

In the more international institutional environment, sovereign nations are supportive of the Bank as well. The Bank is the principal organization capable of financing the development aspirations of nations at the most affordable terms. The cash-flow needs of a country, however, often divert the attention of government officials away from the actual economic impact of the project. Over time, the Bank is viewed first as a source of needed cash (the main concern of finance officials), and only then as a means to promote development. The actual impact of the financed projects becomes a tertiary consideration. The Bank has often excused its deficiencies by emphasizing that it only finances projects, not implements them. Or pleading national sovereignty, it explains that the loans provided were at the government's request, leaving the Bank powerless to do much else.[9] It also claims to no longer finance those types of projects that fail, having moved on to a new, more promising technical package. The need for cash and the desire to move as much money as it can has often resulted in the Bank's financing second phases of projects even when the first phases failed to meet technical objectives. For both the Bank and the borrower, the original project has created a structure through which more resources can be easily passed. This common occurrence, perhaps more than any other, confirms that both the Bank and finance officials have been more interested in the movement of money than in other objectives.

Another reason for the Bank's institutional success is that it has become an important symbol for the Western industrialized countries, in particular the United States during the Cold War. For the U.S. and many Western nations, the Bank serves as the standard flag bearer for the capitalistic approach to international economic development, a supposedly more efficient way to develop than a communist alternative. Consequently, it has provided an important service that many donor governments have been willing to support, politically as well as financially. It is also a mechanism for the Western financial institutions to expand their investment portfolios by tapping into Third World development efforts and markets.

In short, the Bank has succeeded as an organization because it has performed well within its technical environment by moving money and opening investment opportunities for private and public sector investors, while providing crucial political support and symbolic services within its institutional environment. This is why the Bank has continued to grow and prosper, even while the negative economic, environmental, and social impacts of many of its projects were continually called into question at the microlevel. The 1990s, however, appear somewhat bleak for the Bank, since its lack of attention to impacts at the microlevel, as well as the collapse of a communist threat, may be finally catching up with it.

FAO: Strong Institutional Environment

Figure 6.1 shows that FAO Forestry exists within a weak technical, but strong, institutional environment. The literature suggests that inputs and outputs from organizations found in such environments tend *not* to be tightly linked. Supporting and adhering to established norms, beliefs, expectations, and so forth are equally, if not more, important than achieving technical objectives.

I would modify this a bit by saying that organizations in institutional environments are more likely to receive inputs if their outputs are responsive to the demands of elements from that environment. The key point is that the demands are often not simply technically oriented. Therefore, the outputs of organizations in institutional environments tend to be varied, even conflicting and not directly linked to the inputs.

FAO Forestry, along with a number of international agencies, was created to serve the varied needs of the world community, as a collection of sovereign nations. Or stated another way, FAO Forestry was invented by the elements of the institutional task environments to serve them. Unlike the Bank, however, it was not given the capabilities to create its own input resources. As a result, FAO forestry's task environments have considerable control over both the organization's inputs as well as its outputs. This makes FAO Forestry a severely constrained organization. It is a service agency geared to fulfill member governments forestry needs, not unlike a public school that is held accountable for all the varied educational needs of a local community.

As a service agency, it must not offend its members. Although technical professionalism within the organization is important, it ultimately must take a back seat to diplomatic concerns of normative behavior. Its principle defensive strategy is a weak offensive one. To minimize the controlling influence of its task environment as much as possible, FAO Forestry attempts to shape the demands and support of its task environment members with the use of its unsurpassed professional knowledge and information, while at the same time attempting to fulfill their wishes. Conferences, meetings, and publications are the key strategic tools of such organizations for rallying its task environment members, in a consensus-building format, around the activities it wishes to pursue. It also requires professionals with diplomatic acumen to successfully negotiate these affairs.

FAO Forestry is only capable of instituting this strategy if it maintains its legitimacy as an institutional organization. Here it must follow established rules and norms. In particular, it must acknowledge that it is a service organization, that there is equality among all members, and that the members' sovereignty is a fundamental right. To enhance that image, an international staff representing countries from all world regions is assembled. FAO Forestry also consciously attempts to distribute its work and attention fairly evenly among world regions, if not countries.

It must be noted that simply because an organization exists in an institutional environment, it does not have to be technically ineffectual. The constraints placed upon it makes consistent high technical performance more difficult, but not impossible. *Although other concerns may draw the elements in FAO's task environment away from allowing it to fulfill a project's technical objectives, it is also quite possible for elements in the environment to demand technical competence as well.* Simply put, donor as well as recipient governments can demand technical competence. This helps explain FAO's highly variable performance.

Similarly, performance in the institutional environment may not, and perhaps should not, always mean the simple achievement of technical competence. For example, performance to a recipient government forestry agency may mean the opportunity to experiment with new extension structures and procedures. Although the technical results of the experiment may be poor—with, for example, few trees actually planted by farmers, and the foresters turn out to be poor extension agents—the project itself might serve as an important training tool, with inexperienced forestry staff becoming exposed to the concepts of community forestry and extension practices. Again, the question of performance for whom becomes a key issue. With FAO, it is essential to remember that the primary target population of its services has been the recipient government's forestry department, not necessarily poor communities.

Although organizations in institutional environments may be constrained in their ability to perform well technically, they do possess some advantages. The principal advantage is that such organizations have the legitimacy to work fully within the institutional environment to attempt to influence it. Although the technical work of FAO may be controlled largely by elements in its institutional task environment, and the advice it offers to governmental officials may not be fully heeded, it has the legitimacy to offer that advice and work with government officials directly to promote change. Organizations found within technical environments, like CARE, do not.

Institutional Theory, Isomorphism, Organizational Fields, and Generalizing Results

Instead of attempting to explain organizational differences, the institutional theory of organizations (Meyer and Scott 1983; Powell and DiMaggio 1991) tries to explain why many organizations are similar. The theory holds that when organizations face very similar environmental pressures, their structures will begin to resemble each other over time, through isomorphic processes (DiMaggio and Powell, 1983). Organizational fields represent a grouping of similar organizational types, where isomorphic influences will be the strongest.

As was noted in the Introduction, I was expecting to find significant differences in the dates when the three organizations started their community forestry programs. Instead, the differences were trivial. Each of the three organizations adopted community forestry programs at roughly the same time. Clearly, isomorphic processes of new institutionalism of some kind were at work (DiMaggio and Powell 1983, 1991). As it turned out, however, the homogeneity among the three organizational programs was superficial at best. The characteristics and outcomes of these three programs could not have been more varied. How can we explain both the superficial similarity and the substantial differences?

The differences partly have to do with the differing configurations of their institutional-technical environments, just discussed above. And noted above, greater conformity would likely be found among organizations within the same organizational field. It is easy to see that the three organizations are each in a different field, with the Bank an international financial organization, FAO an international technical assistance organization, and CARE an international PVO. One could easily predict that greater conformity would exist among organizations that have more similar internal elements and missions and that face nearly identical external forces.

From this discussion, however, arises a puzzling question. If the organizations are in different organizational fields and appear to have different technical/institutional environmental mixes, then how do we explain the adoption of community forestry programs at virtually the same time? If the views of institutionalism and the processes of isomorphism are correct, then it would seem that the three organizations must exist together within a broader organizational field as well, in addition to the more specialized or narrowly defined fields already discussed. If so, this leads to a notion of nested levels of organizational fields. I think that although the three organizations are located within different, specialized organizational fields, they also all exist within a more general one: international development assistance (or perhaps, rural development). Here the more macrolevel, societal norms have been able to influence and penetrate these very different organizations. At the same time, aspects of their community forestry programs have easily succumbed to the influence of more specialized internal and external interactions. Strang and Meyer (1993) have insights to add to our understanding by focusing attention on how the concept of rural development forestry, as an abstract category, was diffused and institutionalized within organizations that share the same communication networks.

This leads to a second consideration. Would it not be the case that greater success of homogenizing forces would likely depend on the similarity of internal conditions as well as external pressures faced by the organization? Is it not the case that the organization's original mission and internal characteristics, which emerged at its creation, would be central in placing the organization in a particular field? For instance, the similarities among the development banks would revolve around their functions as banks, wherein they loan money to member nations. Consequently, they would have similar internal features as well as external ones. The observation of nested organizational fields provides, then, greater clarity to this comparative case study. The more narrow and specific the organizational field, the more homogeneous the organizations become; the more general the organizational field, the less homogeneous they become. The superficial similarities among the three community forestry programs can be explained by the limited influence of the broader organizational field.

The use of organizational fields provides an important avenue for generalizing this study's findings. Instead of these results simply being limited to the three organizations, the use of organizational fields suggests that these same findings should apply to organizations of the same field. One could apply greater generalizability to tasks as well. Tasks similar to community forestry—those that are socially based and, hence, complex and dynamic in nature—should require flexible technologies and structures in an organization in order for the organization to handle the tasks successfully. Hence, similar organizations pursuing similar tasks are likely to possess similar strengths and weaknesses.

How The Organizations Adopted Community Forestry Programs

Another aspect of institutional theory to explore is the way in which each organization became involved in rural development forestry. The reason each of our organizations adopted this new type of forestry is distinctive (see table 6.3).

The World Bank's adoption of community forestry, for example, was largely the result of a calculated top-down policy decision. The Bank's interest in rural development forestry was folded into its new emphasis on rural poverty alleviation, which developed during the early 1970s. Through the persuasion of his office, and spurred on by the failures from the previous decades of urban-centered, industrial development efforts, Bank president Robert McNamara forcefully guided the Bank into rural development activities. To fulfill McNamara's directive to do more to combat rural poverty, the Bank staff was sent scurrying to develop new programs. Forestry, geared to addressing rural poverty, was one of those programs. The staff reviewed the activities of other international organizations for ideas and sources of technical support. One such find in their search was FAO and its activities in forestry (Brechin 1989, 115). The Bank's adoption was aided as well by its relative autonomy and ability to implement its own policy decisions.

FAO Forestry, on the other hand, was and has remained a heavily constrained service organization. Although no direct evidence was found in FAO Forestry's formal records, other sources confirmed that at least a few individuals within the organization, in the early 1970s, felt the need to move it into community and rural development forestry. The most vocal individual was Jack Westoby of the United Kingdom. Through his action and writings, Westoby became a principal international figure for "people forestry." He worked for FAO Forestry for many years in a number of capacities. While there, he became an early clear voice, calling for a new kind of

Table 6.3
Comparison of "How" the Organizations Adopted Community and Rural Development Forestry

	Movement into Community and Rural Development Forestry
World Bank	Top-down policy decision, McNamara directive, Midlevel mimickry
FAO Forestry	Interagency collaboration: SIDA-FAO agreement, 1978 World Forestry Congress, International Legitimization, 1979 World Conference on Agrarian Reform and Rural Development
CARE USA	Bottom-up project activities (local government forester and villagers) and decentralized CARE mission decision-making ability

forestry to meet the needs of poor rural communities. Westoby left FAO Forestry in 1975, frustrated by the failure of the international community to institutionalize this new kind of forestry. He felt he could be more effective working from outside rather than inside the organization. He left FAO Forestry before Sweden began negotiations with FAO in 1976 to begin supporting the agency's shift into people forestry (Arnold 1988, Westoby 1975, 1987; personal communications with Westoby 1988). Working from the United Kingdom, Westoby became an important force behind the 1978 World Forestry Congress in Jakarta, which brought the need for people forestry to the international stage.

Eventually, with the political and financial assistance of the Swedish government, FAO Forestry began to develop its expertise in community forestry as a part of its Regular Programme. Of interest to us, FAO Forestry, owing to its political and financial constraints, was unable to adopt community and rural development forestry on its own. It needed the assistance of another organization, a member nation, and the legitimacy derived from two international conferences on the subject before the organization could become formally active.

Sweden, through its trust fund support and legitimacy as a member of FAO, played a vital role in allowing FAO Forestry to quietly build up its own expertise on the subject, with the whole affair framed simply as the organization's serving the wishes of a member nation. Not until the World Forestry Congress in 1978, and FAO's WCARRD in 1979, did international political consensus finally formally institutionalize the value of community and rural development forestry.

CARE, on the other hand, became involved in rural development forestry because of more spontaneous, community-level, bottom-up actions. Because of the severe environmental problems they faced, local farmers in central Niger approached the district's forestry official and his Peace Corps Volunteers assistant for help. They developed a windbreak scheme to protect the relatively fertile agricultural valley from further degradation. The forester and his Peace Corps assistant then approached CARE to see if they would back the idea. CARE's nongovernmental character and established presence in the rural areas of Niger made CARE accessible and responsive. Likewise, working from its permanent country missions in Niamey, CARE's Niger field staff was constantly searching for new program ideas. Because of this and its own internal source of finances, CARE was able to respond quite quickly to the request. The result was the Majjia Valley Windbreak, which became the most famous agroforestry project in Sahelian Africa. CARE's work in community and rural development forestry quickly gained momentum as the need for this type of forestry continued to be realized by CARE and a number of funding agencies, especially USAID.

In summary, drawing upon theoretical arguments from institutional theory in sociology (DiMaggio and Powell 1983, 1991; Strang and Meyer 1993) as well as political science (Hall 1993), our consideration of how these organizations came to adopt community forestry practices leads to several insights about how organizations in general institutionalize new practices. Following the general lines of argument presented by Hall, community forestry, as part of broader approach of rural development, replaced former policy instruments associated with more traditional forms of economic development, that is, industrial modernization. The failure of that "policy paradigm" to explain the vast sea of rural poverty led to the development of the new paradigm of rural development. In addition, in Strang and Meyer's terms, the cultural links associated with the "theorization" of rural development emerged, and was diffused among development institutions as a new "compelling model of behavior" and as progressive policy (Strang and Meyer 1993, 499).

Adapting DiMaggio and Powell's (1983, 1991) notions of institutional isomorphism, or why organizations have similar structures, helps us, although only partially, to frame our understanding of precisely how the organizations adopted rural development forestry innovations. DiMaggio and Powell (1983) present three basic ways in which organizations homogenize. First, they can mimic, or copy, from each other. A second way is through coercive pressures, largely from actions by government or powerful organizations from within their fields. The third is from normative pressures, largely the consequences of professionalism, most likely from the transfer of personnel. Our cases seem to be, in part, a strange mixture of all three types, but in some ways are devoid of them as well. These examples offer a brief but textured analysis of how these organizations adopted community forestry practices. They are each, in sum, the result of internal as well as external factors, and global as well as local aspects of the change processes.

What is also interesting is that the adoption of community forestry innovation came from different directions. For the Bank, the quest for poverty alleviation programs was sparked by the top-down leadership of its president. Here the coercion was not external, but internal. McNamara more or less demanded action, but did not dictate the exact form of that action. The adoption of rural development forestry itself was the result of the Bank staff's copying, or mimicking, the form from FAO. Consequently, the Bank's adoption was the result of a unique combination of internal coercion and mimicking the structures of other organizations.

CARE's adoption came up from the bottom. The need for rural development forestry was expressed first by local farmers and professionals in the field who witnessed the growing degradation of the biophysical environment. Local reality ignited this innovation for CARE. Certainly, the Majjia Valley Windbreak and Guatemala tree-planting projects found a warm home at CARE headquarters where the groundwork for this type of project had been laid by discussions found within the larger environment mostly through "theorization" (Strang and Meyer 1993) based upon the now-obvious need and the emergence of people forestry programs as progressive policy. The closest mechanism for CARE's case is normative, since mimicry and coercion seem inappropriate, but this itself is not totally convincing. More generally, the adoption of new ideas, especially from the field, is very reflective of CARE's organizational character. In short, adopting innovations is part of CARE's organizational approach. This explanation finds some credence in that it was not until 1981, six years after the onset of the Majjia project, that CARE institutionalized the community forestry program at its New York headquarters. This suggests that only after other bottom-up forestry needs were identified from other countries did CARE formally institutionalize this type of program. This incorporation occurred only after the concept had diffused and been accepted elsewhere—primarily at USAID, and by the general public, the two biggest funders of CARE's work in this area.

If the Bank's diffusion is top-down and CARE's bottom-up, the best way to characterize FAO Forestry's adoption is "sideways." FAO Forestry's adoption of community forestry programs came from other organizations, in fact from FAO's task environment. Mimicry and normative approaches do not apply straightforwardly, however. The closest is likely coercion. Although the adoption was not coerced in any formal way, FAO Forestry could not have refuse it either, given the nature of its relationship with its task environment. Of interest here is FAO Forestry's actual willingness to have a community forestry program, although it could not institute such a program until it was assisted by other organizations.

As with the other two cases, FAO's adoption was made possible by the coalescing of several activities on at least two different levels. It would appear that two external circumstances were instrumental in FAO Forestry's adopting a program that it was unable to implement on its own. The first circumstances consisted of the international conferences and the second was SIDA's intervention a few years earlier. The conferences provided the larger political authority for FAO to pursue these types of activities, while SIDA's intervention actually instituted the means and skills necessary for FAO Forestry to actively carry out the new type of forestry. Both allowed FAO to pursue an innovation that was desired internally but that was blocked by protocol until the concepts were formally legitimized by international meetings, a collection of both experts and public officials as suggested by Hall (1993) to make policy shifts.

In sum, the precise adoption processes followed by each of the organizations was unique. Drawing upon the more homogeneous diffusion process at the macrolevel, each organization's adoption was shaped by the particulars of the organization's more intimate interactions with other organizations and its own internal characteristics. As the state may be more relevant in the formulating of shifts in "policy paradigms" as opposed to particular policy instruments or their incremental adjustments (Hall 1993), the nature of specific organizations may be more relevant in the actual adoption of that paradigm.

Organizational Technologies

Technology most frequently represents the work performed by an organization. Scott (1992) notes that organizational theorists have come to include in the term technology "not only the hardware used in performing work but also the skills and knowledge of workers, and even the characteristics of the objects, inputs and outputs, on which work is performed" (227). Perrow (1986) more succinctly, but with even less precision, defines technology as the "techniques or tasks utilized in organizations" (141).[10]

The lack of a standard definition for technology has caused some serious problems in using organizational technology as a predictive variable. Defining technology has been difficult since a precise measurement of technology has proven elusive.[11] Because of these reasons, I use technology more as a sensitizing concept in this study than as a real variable.

Yet, in spite of these problems, technological perspectives have proven to be extremely important for gaining insight into organizations and their performance. This study certainly supports that observation. Perrow (1986), both fan and critic of technology as a variable, remarks that technology, as part of contingency theory, provides one of the best frameworks for understanding organizations devised to date (146). Scott (1987) also believes in the utility of using internal elements, such as technology, to better understand organizations, but he too stresses the need for researchers to be clear in what they mean by such elements (88). With the exception of Perrow (e.g., 1984) technological perspectives within organizations no longer receive the attention they deserve.[12]

In developing a more precise concept of technology for this study, it was important to separate the task and organizational output from the skills, procedures, and tools utilized by the organization to produce the output to fulfill the task. To highlight this separation, the concept of *core technology* was used (see Thompson 1967, 19). Core technologies, then, represent most accurately the organization's ability (tools, procedures, approaches, skills of members) for transforming inputs into outputs. Organizational outputs here were community forestry projects. Consequently, it was the characteristics of the core technologies, structure, and outputs, given the task of rural community development, that were compared across the study's three organizations. Failure to separate the task from technique would incorrectly suggest that the three organizations have the same technology, and thus we would lose an important explanation for the stark differences in the outputs from each organization (see table 6.4, below).

As a result, the characteristics of the outputs, task, as well as core technologies, are very important to this study. Performance depends very much upon the interrelationship among the three. This is a point I want to emphasize, while also acknowledging that it is an elaboration of Tendler's (1975). In Tendler's study of development assistance organizations, she found that the matching processes between tools and task affected the organizations' level of performance (see also Hoben 1989). For this study, I add greater specificity to these components. The possibilities of "mismatches" have obvious implications for organizational behavior and performance. Of course

Table 6.4
Separating Technology from Output and Task

Core Technology	*Organizational Output*	*Organizational Task*
transformation process	community forestry projects/activities "bankable" projects	rural community development; poverty alleviation

these are contingency theory arguments. From this research, however, it was clear the output has characteristics independent of those of the microlevel environment, where the output becomes embedded. For example, outputs like community forestry projects have a number of technical concerns—types of trees to plant, their planting and growth requirements, planting configuration, growth rates, economic viability, a framework for implementation, and so forth. Although these are important and complicated concerns, what makes community forestry particularly difficult is the unpredictable social, economic, and political context in which it is placed. Issues of tenure and equity, community organization, government regulations and relations, cultural differences, and human perceptions make a relatively simple technical activity into a monumental undertaking.

Empirical Evidence and Thompson's Technology Types

Regarding transformation or core technologies, Thompson (1967, chap. 2) listed three basic types: long-linked, mediating, and intensive. Perrow (1967) constructed a very similar typology. A *long-linked technology* is the simplest of the three and is like an assembly line in character. Here work is performed in sequential order where A occurs before B that occurs before C, and so on. Uniform inputs become assembled into uniform outputs. *Mediating technology* attempts to represent a standardizing process. This type of technology transforms diverse inputs into standardized or uniform outputs. *Intensive technology* is the most complex. This typology suggests a customizing transformation process in which diverse inputs are shaped into diverse or customized outputs. For this technological type, effective feedback activities are essential during the transformation process.

This discussion raises the issue of the possible utility of Thompson's conceptualization—in particular, his distinction between mediating and intensive technologies. Clearly, the Bank utilizes more of a mediating technology in attempting to take complex and uncertain information and create a more standard package, for example, a bankable project. Bank appraisal documents necessitate considerable amounts of detailed information (financial, economic, technical, etc.), to make the standardization process work. For the Bank, at least, this also has made the output itself rigid. Consequently, both the organizational output and the community development task have been held "hostage" to the Bank's core technology, and the output has not been shaped the task. The result has been poorer performance for this task. FAO Forestry's Field Programme and CARE, on the other hand, have possessed intensive technologies. Here, their more flexible and reflective technologies and outputs have been more easily adjusted to fit the task, thus improving performance.

Clearly the transformation process can affect the characteristics of the outputs. When the community forestry output became a bankable project for the Bank, it also became an output less attractive to many government officials (e.g., finance ministers), a key element in the Bank's task environment. The concept of a bankable project output, however, has placed the Bank's work in a somewhat different light. Unlike the outputs of FAO and CARE, the Bank's output has harbored several purposes. One purpose has been as an investment package for the ministry of finance to improve the country's cash flow. Another has been as a means for stimulating greater economic growth. Yet another has been as a security, which the Bank owns and may sell in the financial markets to raise more capital to allow it to make more loans. Consequently, the creation of project loans and their agreement with governments are ends in themselves, enabling the Bank to move its money. The actual impacts of the loans themselves can become of less interest to the Bank—as well the borrowing government. This is all rather like the tobacco industry, which sells cigarettes for a profit to customers who want and demand them but pays little or no regard to the consequences. Similarly, the Bank (at least for the greater part of the study period) tended to worry little about the actual impact of the product on the customer, as long as customers still demanded the product. Both industries have denied any responsibility for how their product is used. As the Bank itself has stressed, it is not an implementing agency. But as I have argued, although implemented by others, the characteristics of the Bank's design and appraisal process has had enormous impacts on project results.

Buffering the Technical Core and Performance

One of the central notions related to technology, one which is relevant to my study, is that of performance. From a rational systems perspective, Thompson (1967) suggested that organizations attempt to maintain performance by sealing off their technical cores from environmental disturbances through the use of "buffering strategies." Where strategies are nonexistent or only partially effective, environmental influences can be viewed as *constraints,* affecting the performance of the organization's technical core (Thompson 1967, 22). We would expect differing organizations to have differing technologies and environmental influences, thereby facing different constraints. Consequently, different organizations with different constraints can be expected to have different levels of performance when the same task is compared.

Thompson noted that organizations with different technologies pursue different strategies in protecting their technical cores. More specifically, organizations with mediating technologies have attempted to minimize the risk to themselves by increasing and diversifying their customers. The Bank's very active work to identify new investment opportunities and to stimulate demand for more projects through the publication of sector policy statements, the creation of investment teams, and the establishment cooperative relationships with other agencies, strongly supports that observation.

Thompson also suggested that organizations like FAO Forestry and CARE with intensive technologies attempt to gain control over the task to maintain performance. CARE has certainly been able to do that quite well, while FAO Forestry often has not. The variation in CARE and FAO's behavior and performance can be explained more by their differing levels of control over their task environments than by differences in their core technologies. The Bank, however, given this task, seems to be influenced heavily by the characteristics and constraints of its own core technology, much more so than with the other two organizations.

Organizational Structures and Technologies

Regarding the relationship between structure and technology, Scott (1992), drawing from the larger literature, outlines three general principles, all assuming rationality (231). The first is: The greater the technical complexity, the greater the structural complexity or differentiation or specialization. The second is: The greater the technical uncertainty, the less the formalization and centralization. And finally: The greater the technical interdependence, the more resources that must be devoted to coordination.

There is, of course, a major exception to the first principle. Instead of a more differentiated structure to perform complex tasks, many organizations hire more complex performers, that is, professionals. This is certainly the case regarding the three organizations. Each relies on professionals (foresters, economists, development planning specialists, etc.) to do the transformation. But how was the professionals' work structured? How much autonomy were they given, and where were they located? Professionals with limited administrative autonomy, subjected routinely to close supervision, exist in what Scott (1992) calls heteronomous professional organizations, while the opposite are known as autonomous professional organizations (253). This study's three organizations varied on this dimension, with the Bank the most heteronomous and CARE the most autonomous. FAO Forestry, interestingly, could be either, depending on the specific circumstances of the project and on which of its dual structures was in play.

On this issue of structure, the dimensions of centralization versus decentralization of work and the degree of formalization had some relevance to this study. Centralization in this study referred to actual location of the organizations' professional staff. A centralized structure for this study was one in which a majority of the staff was located at headquarters, while a decentralized one was the opposite. Formalization represents the degree to which established procedures determine the roles and relationships of the positions found within the organization. Greater formalization suggests more precise rules regulating relationships. Hierarchic, bureaucratic ("tall") organizations tend to be centralized and formal. Decentralized ("flat") organizations tend to be less formal. It is important to note here, however, that there is disagreement in the literature regarding these principles. For example, some decentralized structures are found to have more formal relationships. According to Blau and Schoenherr (1971), this inconsistency results from viewing centralization and formalization as possible alternative methods of administrative control. Scott (1992) argues that these generalizations tend to hold for those organizations that maintain high levels of performance and that these generalizations are the basis of contingency theory (262).

A related dimension to the degree of professional autonomy and of decentralization is a concept I have called *technological flexibility.* For complex, unpredictable tasks like community forestry, one needs to have technical outputs, or products, that themselves are flexible and a core technology capable of producing them. This suggests that such outputs and tasks be completed by autonomous professionals to improve possibilities for performance. With regard to organizational fit to task, Tendler (1975) attempts to explain a similar notion as "decentralized professionals working out problems themselves with easy access to superiors" (12). This discussion of technology and structure prepares the stage for our review of contingency theory.

Contingency Theory: Classical and Current Perspectives

As was suggested above, the interconnectedness between internal organizational elements (technology and structure) and the environment has made it difficult to measure those elements precisely and consistently. Still, research on their interactions has given us rich insights and a conceptual framework for better understanding organizations and their performance. The work of Burns and Stalker (1961), Galbraith (1973, 1977), Perrow (1967), Thompson (1967), and especially Lawrence and Lorsch (1967) has led to the development of "contingency theory," which Perrow (1986) broadly calls the technical school of organizational sociology.

The theory reflects the notion that there is no one best way to organize the work of an organization (Pennings 1992). Rather, the best way depends or is contingent upon the environmental demands placed upon the organization and the technologies utilized. In sum, under conditions of rationality, structures are adjusted to deal with uncertainties, whether internally from core technologies or externally from environmental conditions.[13]

Contingency theory is often stated in a number of ways, but one simple presentation is that tall, or hierarchic, organizational structures (i.e., mechanistic organizations) are most effective if they have routine technologies and are found in stable environments. Likewise, organic organizations maintain performance in turbulent environments and have nonroutine technologies. These flat, or decentralized organizational structures perform best under conditions of uncertainty—as long as an effective coordinating mechanism among the units is established (Lawrence and Lorsch 1967). This latter form is often more effective since it can more easily manage the needed communications, information, knowledge, and critical skills required to address complex or difficult technologies. It is suggested that other combinations, such as hierarchic structures with nonroutine technologies or routine technologies with flat structures, are less effective.

Other scholars such as Perrow (1986) and Pennings (1992) also discuss contingency theory, from the viewpoint of, not simply core technologies, but the characteristics of the tasks. This too makes good sense since a nonroutine task requires nonroutine technologies and decentralized structures, assuming a rational system. Overall, the theory suggests that organizational performance is maintained when there exists a congruence, or good match or fit, between the organizational unit and the environment in which it works.

I believe contingency theory helps explain the differences in organizational performances among the three organizations, though it is by no means the whole explanation. To improve performance, a complex and difficult task like community forestry should be matched with a nonroutine technology and a decentralized organizational structure. Two fourfold tables are presented in figure 6.2 (a and b), below, which utilizes task and technology as key variables.

From figure 6.2a we see that the Bank's poor performance is the result of a nonroutine task combined with a tall structure. CARE's success, however, is supported by the nonroutine community forestry task with its decentralized, country mission structures. Both examples support contingency theory. Problems emerge, however, in FAO's case. FAO with its nonroutine task and flat structure should have performed as well as CARE, yet its performance was actually highly variable. Nor does contingency theory tell us why the Bank maintains a tall structure for a nonroutine task.

From figure 6.2b we note that both CARE and FAO have nonroutine (intensive) technologies. But the Bank has a more routine (mediating) technology, characterized in this way because of the inflexible quality of the Bank's need to produce standardized bankable projects. Again, ideally, performance is enhanced when routine technologies are coupled with tall structures and face a stable organizational environment or task. Likewise, performance is maintained when a nonroutine technology is augmented by a flat structure placed in a turbulent environment. Here again we find FAO and CARE together as above, with the same questions relevant. Contingency theory would suggest greater performance for both CARE and FAO.

The unevenness in FAO Forestry's performance suggests that there are other factors that have interfered even when the basic conditions to fulfill contingency theory are present. If this is the case, then contingency theory can only predict the conditions amenable to success, not success itself. With FAO Forestry, its service character and its extreme dependence on other organizations for resources and goodwill have made its level of performance more vulnerable to external influences, and hence more variable.

With the Bank, however, we have a situation where its routine technology and tall structure are mismatched with a complex and unpredictable task located within a turbulent microlevel environment. This combination cannot produce successful performance. Thus, contingency theory seems to explain quite well why the Bank has failed in community forestry. At the same, although the Bank is structured inappropriately for doing a good job in community forestry, its growth, from its inception to at least the mid-1990s, implies that it is performing well overall. This suggest that its structures have been more than adequate for overall survival. This is

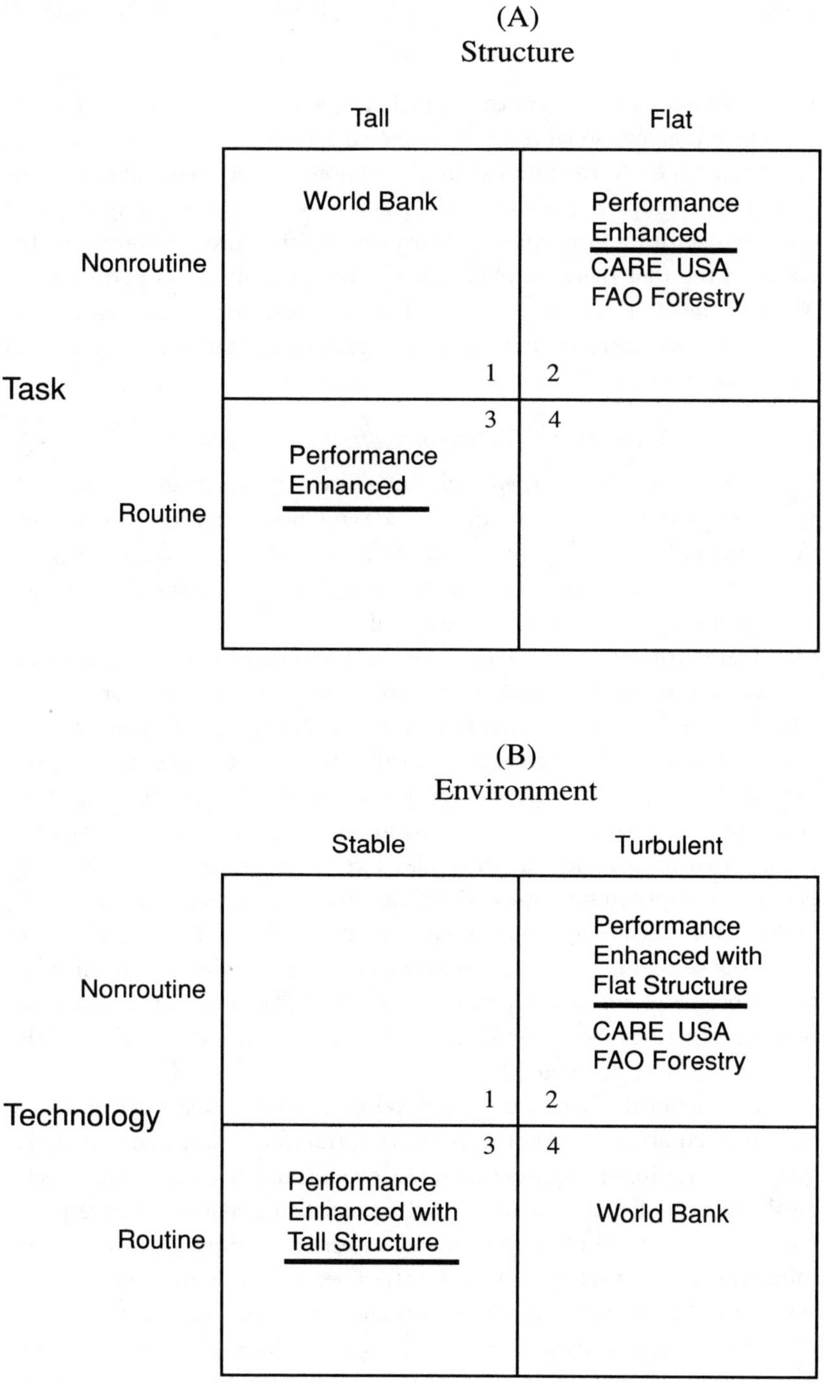

(A)

Structure

Task	Tall	Flat
Nonroutine	World Bank 1	Performance Enhanced CARE USA FAO Forestry 2
Routine	3 Performance Enhanced	4

(B)

Environment

Technology	Stable	Turbulent
Nonroutine	1	Performance Enhanced with Flat Structure CARE USA FAO Forestry 2
Routine	3 Performance Enhanced with Tall Structure	4 World Bank

Fig. 6.2. Contingency Theory

obviously one of the strengths of utilizing a structural-functionalist approach in organizational research. If the structure is prospering when it is functioning poorly, then the essential function has not been identified. Unfortunately, contingency theory does not provide us with insights as to those possibilities. Contingency theory's inability to provide those insights suggests too that there are obviously multiple definitions of performance. What is the clearest from this analysis is that performance is a social construction since there are numerous interpretations that can be applied to this construct for each organization.[14]

Galbraith's View of Contingency Theory

While contingency theory via Lawrence and Lorsch and Perrow provides some insight, it remains quite crude and incomplete in explaining organizational behavior. With the Bank's great ability to generate inappropriately designed projects, Galbraith's "complexity of information" version of contingency theory seems quite useful.

Galbraith (1973, 1977) states that as the complexity of the task, or the environment in which the task is embedded, increases, the organization must process even more information to maintain performance. If these information needs overwhelm the organization (Pennings 1992), the organization has two basic options: Increase capacity to process the information, or decrease performance by processing less information. In addition, given the finding that mechanistic organizations like the Bank have more difficulty handling complex information (Pennings 1992), and given the unlikely case of the Bank's becoming a more organic organization, the Bank would likely select the second option. Slack resources refers to a reduction in information processing, and hence performance, and tends to occur in less competitive situations (Scott 1992, 35). These are precisely the conditions the Bank has historically enjoyed.

Because of the Bank's need to develop complete project designs prior to implementation as part of its mediating transformation technology, the success of the project requires the collection and analysis of all relevant technical, social, political, economic, and financial information. The requirement to integrate all this complex and uncertain information into a single comprehensive project document has often become too demanding. As discussed in Chapter 1, the Bank designs and appraises a variety of projects. The level of complexity and uncertainty surrounding each of these project types varies enormously. Infrastructure projects may be complex, but are more predictable, whereas social investment projects are extremely uncertain, as well as complex, and often vary greatly from site to site. As a consequence, Bank officials have concentrated more on their chief concerns—the economic, financial, and some technical issues, and less on relevant social-cultural information, which is nearly impossible to manage in a document anyway.

While Galbraith's version of contingency theory might help to explain the Bank's poor performance in community forestry, especially concerning the development of inappropriately designed projects, it does not seem to shed much light on FAO or CARE. Pennings (1992), however, notes that more-organic organizational forms, such as CARE and FAO's field projects, can more easily process information since they depend less upon bureaucratic channels of information flow. In addition, these more-organic forms are more likely to nuture the formation of interdisciplinary teams that will better be able to tackle more complicated tasks.

There may also be distinct features of the core technologies that may require different informational needs. In short, different core technologies may have different informational requirements, not just abilities to process information. The Bank needs much more information up front for its investment-oriented, mediating technical core. FAO and CARE have faced the same uncertain task, but have not had to generate detailed implementation documents prior to execution. They have not had to anticipate and control all aspects of a project form the start; their core technologies and outputs have allowed to be flexible in solving problems as they arise. If so, Galbraith's version of contingency theory may also depend upon the specific type of technology and how it uses information, with mediating technologies having more demanding information loads required for standardization, while intensive technologies much less so. And given its rather uncompetitive situation, the Bank would likely follow a strategy of slack resources with decreased performance for socially unpredictable tasks. This would probably give the Bank problems with most, if not all, socially based tasks, which tend to be by their nature inherently unpredictable, requiring flexibility in design and implementation.

Current Discussions concerning Contingency Theory

More recent discussions of contingency theory have focused on several issues, including the degree of the "free will" of organizations to strategically select their most efficient structural forms (Child 1972, 1975; cf. Pennings 1992) and including contingency theory's relationship to institutional theory (Gupta, Dirsmith, and Fogarty 1994). Contingency theory continually raises the issue of the degree of flexibility that organizations have in determining their structures. By its very nature, contingency theory implies that organizations do have some flexibility in adapting to environmental and technological disturbances; witness the number of strategies outlined by Thompson (1967) to manage such disturbances. As stated by Pennings (1992), certain perspectives such as population ecology perspectives (Hannan and Freeman, 1984, 1989) argue that organizations have very limited ability to undertake meaningful structural change and to select their environments. This structural inertia is powerful and not easily shrugged off. Child (1972, 1975; cf. Pennings 1992) takes a more proactive position, stating that organizations have a "strategic" choice in positioning themselves against environmental conditions. Environment not only produces constraints but resources for exploitation (Pennings 1992). Hence, a debate is under way concerning the amount of discretion that organizations actually have in making basic changes to their structures, which can affect their performance—even their survival.

Although this analysis did not address the issue of organizational survival, the three organizations provide an interesting comparative test of the ability to change structures and shape environmental conditions. As I have discussed at length earlier, CARE has considerable ability to select the environmental conditions most favorable for its performance. FAO has possessed very little since it is almost totally dependent upon its relationships with other organizations for its activities. The Bank falls in between. Its autonomy is fairly high since it has the financial means and legitimacy to implement its own policy initiatives. Only when it is under public scrutiny, or is dealing with a powerful borrower, does the Bank lose some of its flexibility. These examples, however, show a wide range of ability in demonstrating strategic choice. Not surprisingly, however, the level of ability seems to rest largely on the level of external constraints encountered by each organization. More to the point, the constraints seem to reflect the very character of the organizations, with FAO's role as a service agency to sovereign member nations being the clearest example.

Fine-grained analyses would most likely find the processes of institutionalization affecting microlevel structures and activities, as suggested by a number of authors (Gupta, Dirsmith, Fogarty 1994; Pennings 1992; Meyer and Rowan 1977; Scott 1987). In my coarser-grained study, I have used both contingency and institutional theories to more fully understand the behavior and performance of the three organizations. Institutional theory has been helpful in understanding the diffusion and isomorphic adoption of community forestry as a policy or programmatic innovation. Contingency theory, however, has been useful in outlining the basic characteristics that can explain differences in levels of performance. Still, it has been the unique nature of the organizational structures and the particulars of resource dependencies that have enhanced or hindered success and shaped how the community forestry innovations have been adopted. Only by combining a number of organizational perspectives can we hope to more accurately understand organizational behavior.

Emerging Ecological Community of International Forestry Organizations

While researching the role of international organizations in the community forestry revolution, I inadvertently uncovered a second revolution—an organizational one. With the rapidly growing awareness of global deforestation and the need to stimulate sustainable development, an active and distinct organizational community is taking shape around a broadening set of international forestry issues.

Dimensions of an Ecological Community

This community's distinctive geographic boundary is the global system, not the typical local-level community found in the sociological literature.[15] It is also a diverse collection of organizations that is forming its own community, as defined by Astley (1985), but with its own political economy (Benson 1975) and with members attempting to either cooperate with or co-opt one another on certain issues.

This community has formed around an international forestry function, most noticeably the highly conflictual set of goals of development assistance in forest management, economic development, and biological conservation, mostly in LDCs. In some instances it has deployed the use of more specialized organizational roles. The principal roles are project financing, technical innovation and project implementation, political legitimization, and information dissemination. This is not to say that these activities were ignored earlier, nor am I suggesting that a loose collective of individual organizations involved in matters of international forestry was not in existence prior to now. The difference is that each of these organizations performed a number of these roles independently of one another. The community now, however, appears more coherent, tightly coupled, and specialized, although at times still conflictual. Some attempts at limited and broad-based cooperation have occurred, focusing on the particular strengths of the various organizations, in the general direction I suggested in Chapter 5.

Within social forestry, what has increased the tightness or cohesiveness of the community has been the recent move toward role specialization for project financing and implementation. Of interest organizationally is the increased attention given to NGOs. Their general reputation of being both efficient and effective in their work, whether it is project implementation at the community level, monitoring the activities of others, or politicizing issues, is increasing their value and role within the community network.

Regarding project implementation, since the early 1980s, USAID has become increasingly supportive of NGOs in its forestry-related fieldwork. Its Agroforestry Outreach Project in Haiti was an important large-scale experiment of using NGOs specifically for the project's implementation phase. The project quickly turned heads, achieving plantings and survival rates beyond the wildest expectations. It was USAID's first successful forestry project in that country, a rare rural development project success story for Haiti. More important for USAID was that the new arrangement allowed it to relinquish its project implementor's role to NGOs, so that it could concentrate on project funding and supervision, activities the organization is much more proficient at. Overall, the specialization of responsibilities has been a much more efficient way for USAID to conduct its business, although not all PVOs and NGOs are naturals at promoting grassroots development (see Tendler 1982; Van Wicklin 1990).

It was 1981 when USAID and CARE signed their first multimillion-dollar matching grant for forestry. Before then, USAID missions in a number of countries had cofinanced a number of CARE's forestry projects, but this was the first program grant given from the central office to CARE headquarters as umbrella funding for a number of ongoing and new projects to be technically supported from a central location.[16] A second and third multimillion-dollar matching grant agreement with CARE was signed in 1986 and in 1991. In fact, USAID institutionalized its relationships with NGOs since it was given legislative requirements to fund a certain percentage of its resources through these organizations. The 1987 Appropriations Act increased the target percentage for working with NGOs from 17.5 percent to 20 percent (Brechin 1989). Once again, the rationale for this legislation rests on the strong belief that NGOs are effective and efficient vehicles for USAID to dispense its resources. The grant arrangements, in particular, are important vehicles in utilizing NGOs' strengths of flexibility and responsiveness and should become institutionalized.

The U.S. government, however, has not been alone in its belief that NGOs are efficient and effective implementors of development projects. The Canadian government, through the Canadian International Development Agency (CIDA), has had a long history of working through NGOs, including CARE (see Cazier 1964). It too in the late 1980s began its "Africa 2000" policy initiative that was designed to stimulate natural-resource-related development efforts on the continent by funding projects with national and local level NGOs (Brechin 1989).

Bilateral development assistance agencies, however, are not the only funders that have become interested in NGOs. Recently, the World Bank has attempted to work with NGOs. In 1988, the Bank jumped on the NGO bandwagon and began an institutionwide effort to increase its work with these organizations (Beckman 1991), although its success has been very mixed (Beckman 1991; Brechin 1989; Nelson 1991). Much remains to be learned about why some of these arrangements seem to work and why others do not. Can cooperation-coordination failures be explain through structural or systemic problems, or are they simply random and particular? In addition, mapping out the precise arrangements among the different kinds and levels of NGOs and their cooperation with international organizations should be done. How do grassroots, national, and international NGOs relate to one another (see, e.g., Fisher 1993; Peluso, Poffenberger, and Seymour 1990) along with other types of national and international agencies and organizations?

Perhaps the most dramatic effort at tighter coupling within the broader international forestry community was the formulation of the Tropical Forest Action Programme. The TFAP effort provides a fine example of Aldrich and Whetten's (1981) notion of an "action-set" that a group of organizations temporarily forms to pursue a limited purpose. Although TFAP has been enormously controversial and disappointing to many, it was, in my opinion, a glimpse into the future.

TFAP came to life in 1987 when the Bank, FAO, UNDP, and WRI, with some help from the Rockefeller Foundation, jointly created a general framework to address the forest conservation and development challenges of the world's developing countries. The consortium's basic objectives were to stimulate sustainable development and create new jobs and economic opportunity and yet provide for sound and rational management of forest resources, such as fuelwood, timber, water, and so forth, while protecting biological diversity through conservation efforts. Particular attention was to be placed on local community involvement and participation in these nationally based efforts. These goals were to be achieved through increased coordinated development and technical assistance, and investments from international and bilateral development organizations by helping governments create and implement national forestry action plans (NFAP). By 1993, ninety countries and fifteen donor agencies had been involved (Oksanen, Heering, and Cabarle 1993) although the glow of TFAP's promise has now faded for many agencies, including the Bank and WRI, two of the founding members.

It is not my intent here to evaluate TFAP's effectiveness or recap the controversy (see Brechin forthcoming; Lohmann and Colchester 1990; Oksanen, Heering and Cabarle 1993; Winterbottom 1990). but rather to comment briefly and sociologically on the significance of this international effort. Interestingly, this type of coordination effort reflects, more or less, the findings of my study that were suggested in Chapter 5.

Rogers and Whetten (1982) explore interorganizational coordination and cooperation efforts, outline their rationales and pitfalls, and end up a bit skeptical about the outcome of such efforts. The successes and failures of TFAP should be seriously studied and the lessons noted. This is precisely the type of cooperation that will continue to be proposed in the future as institutions attempt to respond to cries for action. Can we find ways to improve and encourage greater cooperation and coordination among international organizations, or should we look to other models?

The organizational members of the international forestry community continue to grow and the collective agenda becomes more active and complicated. The International Tropical Timber Organization (ITTO), located in Japan, emerged out of the 1985 International Tropical Timber Agreement (ITTA). ITTO's central mission is to facilitate the trade in tropical timber between supplier and consumer countries. More recently, it has moved into areas of national policy development, procedural guidelines, and development assistance geared toward promoting sustainable tropical forest management.

The international forestry plot thickened considerably at the 1992 Earth Summit in Rio (UNCED) with the controversy over a forestry convention. Disagreements among nations of the Southern as well as the Northern Hemisphere over signing a binding agreement forced a compromise in putting forward a nonbinding set of forestry management objectives, known as the Forest Principles. Although the Principles uphold rights of national sovereignty, they do recognize the need for greater international cooperation to sustainably manage these resources. This is the first time issues related to the management of the world's forests have reached such international prominence.

And attention continues to mount. The Commission on Sustainable Development (CSD), the body established to oversee the implementation of UNCED's program for solving the world's environmental problems, Agenda 21, in the spring of 1995 formed a World Forestry Panel. The purpose of the panel is to coordinate CSD's work on international forest management concerns. What is interesting sociologically, in addition to the increase in organizational activities, is that these efforts challenge FAO Forestry's organizational domain, an issue that has yet to be fully acknowledged or addressed. Although members of this community have come together on certain issues, such as donor coordination through TFAP, the larger issues of dominance within the international forestry community, traditionally held by FAO, will likely become more contentious in the years ahead.

Forces Shaping the International Forestry Community

Two related developments appear to be behind the recent emergence of the international forestry community. One is greater turbulence within the international forestry setting generally. The second is the desire of a number of major players to respond to the increasing attention as an opportunity to stimulate greater demand for their services. This, not surprisingly, has resulted in even more disturbances.

Until international interest in the biophysical environment arose in the 1970s, followed by continually mounting alarm over the increasing rates of tropical deforestation and the loss of biological diversity, international forestry was an issue soundly lodged in distant backwaters of the international arena, starved for attention. This was despite the constant call for decades by foresters for a greater share of the international spotlight. Organizations involved in international forestry are embedded now in an increasingly complex and turbulent organizational environment. Their organizational environments are rapidly changing because of the accelerated deterioration of the forest resource base and its implications for all living entities and because of the concern over this reality expressed in policy circles. Today the fate of the world's forests are linked to serious environmental issues such as global warming.

Our understanding of these situations has been enhanced largely by improvements in communication technologies. These developments, along with the increased monitoring activities of NGOs, only aid in informing the world community about the human and environmental problems that exist. The growing turbulence over international forest management, deforestation, and sustainable development can be easily traced, in large part, to the activities of environmental NGOs and human rights, and international government organizations like the United Nations Environment Programme. NGOs, as a loosely coupled functional field (group of similar organizations), have garnered legitimacy from research and the use of scientists. Activists, supported by the general public and a few of their politicians, use intensive media coverage, educational programs, publications, and publicity stunts to successfully generate public awareness and concern to a point requiring significant political responses.

A consequence of the combination of the transportation and communication revolutions and these monitoring organizations has been that the biophysical environment is placed under increasing scrutiny. Communication about events in our world are much more commonplace. In short, organizations are more closely inspected now and are held accountable for their failings—witness the U.N. and NATO in Bosnia. With the growth in the number of environmental organizations over the decades and their increasing politicization, a multitiered community of critical environmental organizations are publicly monitoring the success and failings of organizations and nation-states acting in this arena. In particular, the Bank has come under intense fire in recent years for its project failures and lack of environmental sensitivity, largely from the efforts of local and international NGOs (Rich 1994). Adjustments at the international level will continue in large part because of the growing anger and protest at the bottom. Grassroots movements and protests continue to mount (see, e.g., Ghai and Vivian 1994) in response to continued social and biophysical environmental injustices (Broad and Cavanagh 1993; Johnston 1994). Greater international responses will emerge in the coming years to correct these mounting social ills.

When organizations are in the same community (e.g., international forestry and development assistance), they each feel the same pressures of criticism, although the specifics may differ. Consequently, organizations within the international forestry community will likely increasingly turn to one another in an effort to either minimize their weaknesses and maximize their strengths or to at least diffuse the criticism among themselves. Improved performance may reduce the criticism and allow organizations to continue to prosper, but so may spreading the blame among members of an intertangled web. In addition, because of their limitations, individual organizations may have difficulties in affecting the larger environment (as noted by Perrow 1986); but communities, with a collection of skills, increase the likelihood of effecting change that encourages their existence. Thus it was the intent of a number of the TFAP framers to use the opportunity to further promote their services. Ironically, the problems resulting from their effort increased the turbulence, not lessened it. I believe there will be growing environmental pressure on international forestry organizations to become an even more cohesive community.

There is support within the organizational literature for this general conceptualization of environmental change, namely, that open systems evolve from simple to more complex conditions. Emery and Trist (1965) and Terreberry (1968) argue that organizational environments tend to evolve over time from simple to complex and from stable to turbulent because of growing populations and technical advances. In addition, and of interest in this argument, is that organizations tend to respond to more complex and turbulent situations by increasing their departmentalization and specialization (see Scott 1992, 148). Although these notions of increased complexity and specialization were in reference to individual organizations, it seems reasonable to extrapolate these concepts to the ecological community level. Support for the belief in the natural evolution of organizational environments remains controversial in the organizational literature. The activities today surrounding international forestry issues, however, seem to support this theory. It could also be argued that with greater attention and time will come greater calm instead of greater turbulence. This remains to be seen within international forestry. A calm in this domain will only emerge if tropical deforestation somehow becomes a moot point or if something more than a merely adequate organizational solution is actually achieved. Neither is likely to happen any time soon.

Toward a Sociology of International Organizations

As our world continues to become smaller and our collective fates more intertwined, international organizations will take on even greater responsibilities in managing global concerns. At the conclusion of the Earth Summit came the realization that even more development assistance and financial investments are required if we are to keep up with the growing needs of mankind and a rapidly deteriorating biophysical environment. It is a cliche, but a true one, that with the future comes greater challenges. By 2050, the Earth and its institutions will need to care for at least 10 billion people, doubling the present number. Adequately feeding this unbelievable number of mouths is alone likely to result in two and a half to three times more demand for food over our present levels of production, which to some is unlikely to be achievable (Kendall and Primentel 1994). Additional land and increased yields to provide this food (as well as other raw materials) will be essential. Some of that land will have to come from forests. Somehow we must find the means to soundly and rationally provide for these people or face even greater poverty and environmental destruction. In responding to a more interdependent world, I see few choices other than the creation of a more active powerful international system of governance, bringing with it greater responsibilities for international organizations. As noted by sociologist Max Weber some time ago, the major social revolution of the modern world remains largely an organizational one.

Concern over the global environment and the welfare of the human race are only two of the interconnecting themes that international structures will continue to address. Oddly, sociologists have focused little attention on international organizations, their growing interorganizational communities, and the environments in which they function. Here sociology could add much to our understanding of these vital structures, to better evaluate their promise and pitfalls as we all strive toward a much more workable world.

Yet, one can ask: Is there anything in particular that distinguishes international organizations from others? This is a difficult question to answer. There are certainly several obvious differences. The most prominent is that the majority of organizations exist within the domain of a particular state, which has the power to exercise considerable authority over them. Most IGOs are consensual organizations, achieved by agreements among a number of member states. They operate within the entire world community, which lacks the same kind of authority as a nation-state. Yet IGOs are relatively weak organizations and highly dependent upon their member nations for both financial and political support. Still, there are other organizations, such as multinational corporations (MNCs), that operate in the same general environment as IGOs but have far greater autonomy than companies found within a single nation-state (see Sklair 1991). How can we explain this significant difference, as well as the discrepancies found among particular IGOs, such as the Bank and FAO? Obviously the internal nature of the organizations themselves, their particular environmental setting, and their interplay within the macrolevel international environment of cultural norms, political regimes, and realities have a significant effect on their state of affairs and levels of performance. But our ability to answer any of these questions is rudimentary at best. We need to know much more about the particulars if we are to better inform future public policy debates. Until we have a clearer understanding of the nature of the international environment and how it influences the abilities of particular organizations to perform, efforts to establish new organizations, planted within the same organizational environment, will not be any more effective.

This research has shown that different international organizations have particular strengths and weaknesses that greatly affect their ability to achieve certain objectives related to human welfare and our biophysical environment. I have argued that, as a result, some organizations are better equipped to promote certain activities over others. One obvious conclusion from these observations is the need for greater specialization and coordination of organizational activities, within even-tighter organizational communities. Tighter interactions can come by fiat or through promotion of greater marketlike competition among the other organizations. Both approaches, however, will require a much more established and active international system of organizations to oversee. At the very least, organizations should be encouraged to pursue those activities they can do well and to refrain from participating in efforts they are ill equipped to handle. Here, more information about the abilities of organizations is necessary to promote this kind of understanding.

In this study, I have argued for the need to match particular organizational strengths to specific required tasks, though only in the broadest of terms. In sum, much more work and refinement is required to extend our knowledge of particular organizational strengths and weaknesses before we can propose more precise solutions. We need to bring the organization back into both organizational analysis and international studies.

In 1988, Gayl Ness and I attempted to explore international organizations from a sociological perspective, to help bridge the gap between political scientists who have developed rich insights into the workings of the international system but who have tended to ignore (more or less) the role of organizations and the environments that shape them. Sociology and sociologists have much to contribute here. By paying closer attention to international organizations and their surroundings, we might be better able to assure that these organizations provide greater service to the world community. We hope others will also take on the challenge of forging a sociology of international organizations. It will be one of the more important pieces of the public policy debates that will determine the future of our global biophysical environment—and of humankind.

Notes

Introduction

1. For readers unfamiliar with these theories, I offer the following brief explanation. More detailed discussion of these and other notions are found in the final chapter. Institutional theory is at present an active area of research in many fields, such as economics, politics, history, and sociology. Within organizational sociology, institutional theory has focused primarily on how an organization acquires specific rules, norms, structures, and so on, and on the effect that the acquisition has on the organization's activities and behaviors (see DiMaggio and Powell 1991). Contingency theory focuses on the effects that particular characteristics of an organization's structure, technology, and environmental conditions have on performance (see Lawrence and Lorsch, 1967). Similarly, an organization's technical core calls more specific attention to how an organization transforms inputs into outputs. Much literature has focused on how the organization protects its technical core from external disturbances. I argue here that the nature of the technical core fundamentally affects the behavior and structure of an organization (see Thompson 1967). Finally, resource dependency emphasizes that organizations are interdependent and that inequity in interorganizational relationships greatly affects behaviors and activities (see Pfeffer and Salancik 1978).

2. Gupta, Dirsmith, and Fogarty (1994) combine contingency and institutional theories to understand their relationship in defining the structures used in controlling and coordinating work within the U.S. General Accounting Office.

3. TFAP, a collaborative effort of several international organizations, including the World Bank, UNDP, FAO, and World Resources Institute (WRI), along with bilateral development agencies and numerous NGOs, was created in 1987 to coordinate development assistance to Third World countries in order to help those countries manage their tropical forests more rationally. By 1990 severe criticism of the program had surfaced, with claims that it was encouraging more deforestation, not less.

4. The work of Selznick and his followers represents a perspective that focuses on the organizational task, one that DiMaggio and Powell summarized as "old institutionalism." With my analysis, I attempt to integrate the "old" with the "new" institutionalism, which I defined in the text. The differences between the two types

of perspectives have to do in part with differing notions of task environment (i.e., the more direct relationships with other organizations) and more general macro-level influences (i.e., new institutionalism). See DiMaggio and Powell (1991) for a superb discussion of these differences.

5. One exception is Hobens (1989), who completed an excellent and detailed organizational analysis of USAID. Hobens, like Tendler, focused on the match between the institution and its task of development assistance.

6. There are, of course, important exceptions, including work by Strang and Chang (1993) and Finnemore (1993). These authors look at how international organizations help to institutionalize new practices at the global level.

7. I merge the two in this work, although a number of specialists have taken pains to make distinctions among a number of "people forestries." Social forestry generally is the larger, umbrella notion of people forestry, whereas community forestry often represents truly collective efforts by specific local communities. Farm forestry often refers to the tree-planting activities of individuals and households in rural areas, while agroforestry practices usually refer specifically to the spatial arrangements of trees placed among other agricultural land uses.

Chapter 1. Using Trees to Move Money

1. In "The World Bank and Agricultural Development—An Insider's View" (1985), M. Yudelman, a former Bank official, provides an illuminating summary of the Bank's thinking about development in the 1960s and 1970s and the Bank's perceived role in the development process.

2. Not to be ignored is the Bank's role in funding a series of important agricultural development activities in the 1960s that became know collectively as the "Green Revolution." The point remains, however, that these agricultural programs were relatively small when compared with the efforts under McNamara.

3. In addition to the members of the World Bank Group, the International Development Bank and the Inter-American Development Bank are also located in Washington.

4. For convenience, the term *World Bank,* or *Bank,* will be used for both IBRD and IDA. IDA will be used only when I need to distinguish its activities from those of the IBRD's.

5. This reflected the wishes of the U.S. and U.K. delegates. A number of the European and Latin American countries were more interested in the Bank than in the IMF.

6. Judith Tendler (1975) makes this very point for all large aid agencies (see also van de Laar 1980, 237). Other observers as well have made reference to the Bank's need to move money and the biases it creates (see, e.g., Ayres 1983, Hobens 1989, Rich 1994, World Bank 1992b, "Wapenhans Report"). USAID and bilateral aid agencies generally, in contrast. provide grants and not loans. This is a very different process wherein one organization expects to be paid back and the other

does not. This creates profoundly different effects on the two types of organizations and on their relationships with others. The most significant difference is in the selection of projects by host governments. Projects supported by loans tend to emphasize those projects that have direct financial paybacks (e.g., export agriculture).

7. The funds for IDA credits come from a general replenishment from IDA's industrialized members and from transfers of net earnings from IBRD. Only countries with an annual per capita gross national product of $610 or less (in 1990 dollars) can qualify for IDA credits. More than 40 countries were eligible for IDA credits in 1992. The terms of IDA credits, which are made to governments only, have ten-year grace periods (usually with 35- to 40 years of maturities and no interest, but have certain fees (see World Bank annual reports).

8. This emphasis on project benefit-cost analysis, however, has waned a bit in the 1990s, in part because of its many failings and need to worry more about environmental and social considerations. For forestry see World Bank (1991a, 1991b, 1994).

9. The Bank experienced an earlier reorganization twenty years earlier in 1952. The 1972 reorganization reshuffled the Loan and Economic Departments. The Loan Department was responsible for all aspects of the project loans. The Economic Department concentrated on a country's creditworthiness and economic problems and on general economic research (Mason and Asher 1973, 74). Prior to the reorganization, staff from the two departments competed for final control of the project loans. The reorganization streamlined the procedure, creating new units but giving more control over projects to project personnel (see Mason and Asher 1973, 74–79). In essence, this change in organizational structure streamlined the Bank's work as a mover of money.

10. A senior forestry official at the Bank recalled the frustration he experienced with chief policy makers in the developing countries in the 1970s and 1980s when he tried to convince them of the advantages of investing in forestry projects. He quipped that he felt "like a damn Fuller Brush man attempting to peddle his wares." He noted that at the time much more work needed to be done in reaching the policy makers. The development of TFAP by the Bank, FAO, UNDP, and WRI in 1987 was, in spite of its problems, an attempt to stimulate that interest.

11. The control over project selection by government officials may seriously affect not only the type of project loan that is accepted but also the outcome of the project. For example, when interviewing two top Ugandan forestry officials, we discussed their experiences as they negotiated a forestry loan in the mid-1980s. They noted that a conflict had emerged between their ministry of finance and the forestry department. The Bank wanted to finance a particular forestry project, but many of the country's forestry officials were against the project for technical and social reasons. But government officials at the ministry of finance told the forestry officials to stop jeopardizing the project because the former wanted the financial resources that would come with it.

12. For a detailed analysis of the Bank's role in the international debt crisis, see Richard E. Feinberg et al., *Between Two Worlds: The World Bank's Next Decade* (New Brunswick: Transaction Books, 1986).

13. To illustrate the spottiness of the Bank's evaluations, its first evaluation of its forestry loans was published in 1991 (see World Bank 1991b), 13 years after the publication of its 1978 sector policy paper.

14. Chapter 8 of Mason and Asher (1973) provides a well-documented, often humorous account of the development of the Bank's project appraisal activities.

15. With the "Wapenhans Report" (World Bank 1992b) there was considerable discussion about this very point. It became recognized by the Bank that its emphasis on simply making loan agreements was harming the impact of its development projects.

Chapter 2. Of Diplomats and Foresters

1. See COFO, *Forestry and the Environment. Action Proposals on Forestry, National Parks, and Wildlife for the UN Conference on the Human Environment: Secretariat Note* (1972a), 3. See also COFO, *United Nations Conference on the Human Environment* (1972b).

2. See COFO, *Forestry and the Environment,* 4.

3. Cited from COFO, *Report of the Second Session of the Committee on Forestry* (1974), 3. See also COFO *The Role and Participation of Forestry in the World Endeavor for the Conservation of the Environment: Secretariat Note* (1974c).

4. Cited from COFO, *Forestry for Local Community Development: Secretariat Note.* (1976a), i.

5. COFO, *FAO Medium-Term Objectives and Proposals for the Forestry Department's Programme of Work 1976–77: Secretariat Note.* (1974a), 6.

6. The following five technical areas became divisions when FAO was created at the conclusion of World War II: Nutrition and Food Management, Agricultural Production, Fisheries, Forestry and Primary Forestry Products, and Statistics.

7. FAO Forestry had the following names: Forestry and Forestry Products Division (June 1946–August 1951), Forestry Division (September 1951–December 1958), Forestry and Forest Products Division (January 1959–December 1967), and Forestry Department (1970–present) (see Phillips 1981, 142).

8. For more information on the emergence of rural development as an important part of the process of economic development, see the discussion on the rise of rural development under McNamara's World Bank in Chapter 1.

9. This is quoted from COFO, *Progress Made in 1984–85 and Main Features of the Programme of Work in Forestry for 1986–87: Secretariat Note* (1986a), 1; but it is based on the forestry development strategy and priorities endorsed by the Twenty-first Session of the FAO Conference in 1981. For the 1990–91 and 1992–93 biennia, FAO Forestry concentrated more on (1) Tropical Forests Action Programme, (2) Global Forest Resources Assessment, (3) support for the 10th World Forestry Congress (1991), and (4) contributions to the preparation of UNCED in 1992.

10. See Westoby (1975), 208–209 for a short but enlightening critique of the EPTA program.

11. This was part of a general shift of emphasis from Europe to countries of the developing continents found throughout other UN agencies. See, for example, El Razek (1982) regarding the international notion of refugees.

12. A number of individuals closely associated with FAO Forestry noted that there was a generational turnover of personnel during the late 1970s and early 1980s and that the Regular Programme lost exceptional talent and personalities, which have not yet been fully replaced.

13. The exceptions are professionals in other specialized fields. For example, the director of the FTPP has a Ph.D. degree in anthropology.

14. In 1982 there was a policy change that tried to limit the number of continuing appointments (i.e., permanent project staff) and use more short-term consultants. This was an attempt to obtain a high-quality match between expert and project. There has also been an attempt to use more consultants from Third World countries. Some of my informants indicated that although the concept is sound, its practice has been more politically motivated then it should or needed to be, thereby hurting technical performance in the field.

15. As a way of illustration, one informant gave the example of a tropical highland range and forest management problem in Costa Rica. FAO Forestry was able to select a professional from Argentina, which has similar ecology and concerns. USAID, on the other hand, would have been limited to U.S. experts from the arid West, which would have been less appropriate.

16. Until the late 1980s, forestry accounted for about 4–6% of FAO's Regular Programme budget. Since then this percentage has declined. These figures were derived from reviewing a number of volumes of FAO Conference's *Director-General's Programme of Work and Budget.*

17. Although FAO Forestry doesn't have permanent field missions, FAO generally does. They tend to be very small, with one or two professionals serving as liaisons between the organization and the host government. FAO does not administer field projects from these offices, as does CARE.

18. The technical programs cut across FAO Forestry's more formal divisions of Forest Resources, Forest Industries, Policy and Planning Service, and Operation Service (see Brechin 1989, 221–287).

19. For example, the budget for the FAO Forestry Regular Programme in 1984 was $8 million. The FAO Forestry Field Programme was $32 million (see (FAO) Conference 1985b, and FAO Conference 1985c). In 1993, for FAO as a whole, the Field Programme budget was nearly $6 billion, whereas its Regular Programme budget was only $651 million (FAO 1993). Kay, (1980) noted this same trend toward field projects throughout the UN system as early as the 1970s.

20. Until the late 1980s, UNDP was the largest single source of funding for FAO Forestry's field projects. It has been a relatively stable source, but it too has

experienced changes in its funding capabilities over time (see Brechin 1994). For a review of the situation at UNDP during the late 1970s and early 1980s, see C. Lankester's report in "Forestry Planning Newsletter," (1983), no. 7, and FAO Conference (1985c).

21. WCARRD, which set the political stage for FAO's movement into rural development, did not take place until 1979, four years after the Bank, for example, published its policy sector paper on rural development in 1975. The Bank's formal adoption of forestry for rural development did not occur until 1978. FAO Forestry established its Forestry for Rural Development Programme in response to WCARRD in 1980. FAO Forestry's FLCD was formally established in 1979 as a independent entity when it was funded by SIDA. The groundwork for its formal establishment, however, began years earlier at the macrolevel. Once again, this is an example of how institutional change is manifested by the interactions of internal and task environment elements.

Chapter 3. Helping Poor Communities Plant Trees

1. Late 1974 is in CARE's 1975 fiscal year. Even through the formal project agreement had not yet been signed, unrestricted money from the Niger Mission was provided to start the project nursery at this critical time to be in a position to plant seedlings during the next year's rainy season. Formal funding of the project was included in CARE Niger's fiscal year 1976 budget (reference to letter from Jack Soldate, CARE Niger, to Leo Pastore, CARE New York, dated November 25, 1974, Niger Files, CARE Archives). This action by CARE helps demonstrate its flexibility and responsiveness. The Majjia Valley proposal was only formally submitted to CARE in September 1974.

2. FFW is a long-standing development assistance program in which U.S. agricultural surplus is distributed to people in developing countries in return for labor for specific projects.

3. U.S. support of PVOs, mostly through agricultural surpluses and their transportation costs, has paid a number of dividends to both the government and these organizations. Surplus food was the unanticipated consequence of the war effort. At the conclusion of World War II, the agricultural community was still producing unprecedented quantities of food in spite of lessened need. There was considerable political pressure for the U.S. government to purchase the food as surplus rather than penalize the farmers for their exceptional performance during the war. The distribution of surplus foods quickly became an important weapon in the U.S. fight to contain communism. The voluntary organizations, especially CARE, provided an efficient and less-political outlet for these gifts, while liberally distributing American goodwill in the name of humanitarian causes (CARE's work in Tito's Yugoslavia is an appropriate illustration; see Cazier 1964, 241). For the voluntary agencies, the food and its transportation greatly reduced their overhead costs while allowing them to maintain or even expand their scope of operations.

4. For comparison, feeding programs have long dominated CARE's activities. In 1987 feeding programs consisted of 33 percent of CARE's total activities, its lowest total ever. In 1982, the earliest year for which I have similar figures, CARE's feeding programs were 52.6 percent of total development assistance. Self-help projects in 1982 were only 12.7 percent. The 1993 figure of 29 percent is probably low. The more recent annual reports have combined feeding programs with CARE's efforts in public health, masking its work in public health directly (see CARE annual reports for those years).

5. The internationalization of CARE began almost immediately, however, with the creation of CARE Canada in 1946, a fund-raising arm in support of U.S. action, as Canadians too sought efficient outlets to aid in the European relief effort. CARE Canada was incorporated as a Canadian national organization in 1977, as that country's aid program attempted to make more use of NGOs (CARE 1986a, 2).

6. Total revenue for CARE USA in fiscal year 1993 was $451 million, with $228 million, or about 50 percent, in the form of free surplus food commodities and free transportation, mostly from the U.S. government. The value of net fund-raising income becomes proportionally more significant (CARE 1993, 40).

7. This information is drawn from a memo dated February 19, 1987, from the International Employment Office to ANR's Deputy Director, Charles Tapp (Brechin 1989).

8. CARE's disaster relief in Mauritania ended in 1987 without establishing a permanent mission. This was due to the failure of the government and CARE to mutually agree on the programs to be established and, more importantly, on how they were to be run. The government wanted CARE to provide the money but not be involved in actual implementation, a condition CARE does not accept (Brechin 1989, 247).

9. This section draws heavily on CARE (1979a) and CARE (1986a, 4–5).

10. This is true for a number of reasons. It would be difficult for CARE to raise money from the public to aid the relatively wealthy in poor countries or to work in countries considered not impoverished. Perhaps more important, countries not desperate by their poverty are simply not willing to agree to CARE's strict requirements of controlling its resources and projects or waste time on its relatively small financial offerings. In addition, Public Law 480 food is not needed except by the poorest countries.

11. This is not to say that CARE can never affect government policy, at least as it is applied locally. For example, CARE has helped establish local community rights of tree ownership at its Majjia Valley Windbreak Project site in Niger. CARE fought hard to win basic local control over the project's trees, fearing that popular support for the project and its future expansion would otherwise be severely jeopardized (see Delehanty, Thomson, and Hoskins, 1985).

Chapter 4. Patterns in the Forestry Programs

1. Please note that volatile up-and-down nature of the graph is due to the way in which these particular data are presented. The total value of multiyear loans are represented only in the year approved. In reality, the dollar value of the loans would have been dispersed over the years and not given out as a single lump sum payment. What is actually being measured is the dollar value of loan agreements at year of agreement.

2. As we have seen, the Bank's support of rural development forestry began systematically in fiscal year 1978 with the publication of the Bank's original forestry sector policy paper (World Bank 1978). Sources at the Bank told me that efforts were made to place rural development forestry projects in the loan pipeline while work on the sector paper was under way so as to have projects ready to fund when the paper was released.

3. This list of countries originated from an 1980 study by FAO of fuelwood supplies in developing countries. Acute scarcity was defined as an inability to meet minimum requirements even by overexploitation of remaining woodlands (see World Resources Institute 1985, 20).

4. See Leach and Mearns's (1988) discussion of gap theory in firewood need.

5. It is important to note here that this evaluation is most appropriate to FAO Forestry's Field Programme and less to its Regular Programme. Only a handful of field projects were sponsored by FAO's community forestry unit FTPP.

6. In fact, these goals were achieved precisely because the government was not involved. Its absence was an intentional and integral part of the project's design. Overall credit for the project design goes to Harvard anthropologist Gerald Murray, with contributions from consultants Frederick Conway, Glenn Smucker, and Karl Voltaire.

7. For the record, CARE had flirted with tree-planting activities in limited and isolated ways for some time prior to the Majjia Valley project. An unpublished and uncompleted history of CARE by Kaufman (1971) made reference to the support of conservation activities in Lesotho in 1971 (38). The project centered on the development of 849 earth dams and the planting of 2,640,000 trees. It was a government project in which CARE provided food for work and hand tools for the planting of the trees. George Radcliffe, Assistant Executive Director of CARE USA, also noted isolated support for forestry projects before 1974. I know from personal experience of limited tree-planting efforts in Haiti prior to 1974, but they were small components of other projects and were not very successful.

Chapter 5. Examining Organizational Behavior

1. Change within international organizations is the subject of ongoing research I am doing on FAO Forestry and its experience with TFAP and on the Bank's Environment Department.

2. The case of Bank's 1987 loan to Rwanda serves as an illustration. Here, the Bank loan was "captured" by high officials for their own use (Brechin 1989).

3. CARE has removed missions from certain countries. In the 1960s, it left Poland at the request of the Polish government, because it had become a political embarrassment for Moscow; it left Greece after that country had become too wealthy; it suspended its missions because of war, for example, in Chad and in Afghanistan during the early 1980s; and it left Mauritania in the mid-1980s after the government became too uncooperative.

4. It is likely fine for the Bank to fund community and rural development forestry projects. More to the point, it should not design them. Clearly, the Bank might be encouraged to finance community forestry projects that are in demand by governments, especially if those projects are designed by others more capable at that task. The Bank would be more proficient at providing financial resources for specific projects already established and successful but constrained only by lack of capital. For example, PICOP, the Bank's first rural development forestry project ironically illustrates this approach. Small farmers grew trees for a local paper and pulp mill that they depended upon for their raw materials. The project started in 1968 as a successful pilot effort and was expanded under Bank financing in 1974 (see Hyman 1983).

The Bank can be effective in community and rural development forestry if the recipient government is committed to the ideals of community forestry and has the capabilities to implement that kind of project. Nepal provided a good illustration of this case. In 1978, the government in a change of policy denationalized the country's forests, returning greater control over to the local communities and developing community forestry units within its forestry department (World Resources Institute 1985; Winterbottom and Hazlewood 1987). The Nepal government asked the FAO Forestry Department to provide technical assistance to the program, and the World Bank provided the external funding. The Nepal case again represented a fine example of a successful tripartite arrangement between funder, willing recipient government, and technical assistance agency in which the parties nicely divided up the work accordingly to their particular strengths. This model should be emulated and is argued for later below.

5. The controversy surrounding TFAP was raised in Chapter 2. Briefly, spearheaded by FAO, the Bank, UNDP, and WRI, and assisted by donor countries, TFAP was to coordinate and rationalize development assistance directed toward tropical forestry management. It set forth a number of laudable objectives, the more important being decreasing tropical deforestation. A number of critics have claimed that the injection of capital through the program has caused the opposite result—greater deforestation, not less—all at the expense of local people, biological diversity, and sustainability.

Chapter 6. A Theoretical Review

1. Abbott and Snidal (1995) discuss the concept of mesoinstitutions in international politics. Here they are referring to the role international organizations play in the larger international politics setting, between the more amorphous notions of international regimes and the more concrete, but limited, actors of independent nation-states.

2. One can see how performance locally can cause more generalized outcomes. For example, one must wonder what the general impact will be of continued government cutbacks. Further elimination of funding support and services will certainly harm the particular effectiveness an agency could have, thereby fueling the belief in the ineffectual nature of governmental programs, lending even further support for more cutbacks.

3. The power-dependency relationship is based on the work of Emerson (1962). Here, power over an organization is proportionate to the degree of dependency of that organization on another. The degree of power is also affected by the value of the resource and is limited to that one resource. Therefore, it is possible for the same organization to be in a powerful position in one situation, yet dependent in another.

4. Other researchers have made similar observations but use different labels. See Thompson (1967), Wamsley and Zald (1973), Zald (1970).

5. Dobbin (1994a, 1994b) makes a strong case that technical environments don't really exist because economic rationality is nothing more than a social construction, as are the elements found within institutional environments. Still, I think there is considerable utility in maintaining the notion of a technical environment, at least as a sensitizing variable. Although economic rationality may largely be defined by a host of ever-changing values and norms, in organizational analysis, at least, I think it is clear that survival for a number of organizations largely depends upon a very close connection of organizational outputs to inputs, which is reflected in the technical environment concept.

6. As noted in Chapter 3, CARE has received much support from the U.S. government (especially food aid and various program grants), yet CARE maintains its autonomy and flexibility by having its own source of income that it generates from sophisticated marketing and donor-relation practices. Competition with other PVOs for donation dollars, from both the public and government, forces CARE to mind efficiency issues.

7. The near future may be rough on the Bank. Many experts and Bank employees are predicting the Bank will undergo a major realignment, which likely will reduce the size of the Bank staff and operations significantly. I am presently exploring the causes and likely consequences of this impending change.

8. As I explained earlier, the Bank raises the funds it uses in the form of loans from the Western financial markets, where it also sells its loans as securities, with the proceeds reinvested as new loans used to continue its operations and growth.

9. In the 1990s, because of considerable criticisms of its lending activities, the Bank has been making greater use of loan conditions when the projects affect the natural resources and the environment.

10. For example, works by Litwak (1961), Rushing (1968), Perrow (1970), Thompson (1967), and Dornbusch and Scott (1975) have looked at the characteristics of inputs as measures of technology. Harvey (1968), Lawrence and Lorsch (1967) Pugh et al. (1969), and Thompson (1967), have used the characteristics of organizational outputs as measures. More in keeping with my approach, Amber and Amber (1962), Huge and Aiken (1969), Galbraith (1973), Lynch (1974), Perrow (1970), Pugh et al. (1969), Rackham and Woodward (1970), Thompson (1967), Udy (1959b), Van de Ven and Delbeeg (1974), and Woodward (1965) have used various characteristics of the transformation process as measures of technology. A number of these researchers used different measures within the same study.

11. See Perrow (1986, 143–146 and Scott 1987, 223–32 for a detailed explanation of some of these problems.

12. Of course, I am ignoring the growing literature on the organization of risk (e.g., the works of Lee Clarke, William Short Jr., and others) since these works focus more on organizational decision making and response to risky technologies existing in the environment (i.e., organizing risks) as opposed to technology as an internal component to or an output of an organization.

13. The term *contingency theory* was first used by Lawrence and Lorsch (1967) in their influential work. Others, as noted above, made similar arguments about the same time. It is important to note that in Lawrence and Lorsch's work their unit of analysis tended to be departments and not entire organizations, since departments within the same organization may face completely different environmental forces. Work-unit contingency analysis remains the most fruitful (see Pennings 1992).

14. The notion of performance is central to contingency theory. It is precisely what is maximized when the proper fit between structure and environment is made. Originally, we defined performance as the design and implementation of rural development projects that provided sufficient benefits for the members of needy, rural communities. When such a definition is applied, this study's findings are upheld. However, this definition may be a bit too narrow when one reflects on the overall offerings of each organization.

15. This is referred to as an "areal" organizational field (see Scott 1992, 128) or ecological community (based upon Hawley 1950).

16. Although this may have been the first USAID grant of its kind for forestry, at least to CARE, similar ones appear to have been made previously to other sectors, such as population planning.

Bibliography

Published Works

Abbott, K. W., and D. Snidal (1995) "Mesoinstitutions in Internatinal Politics." Paper prepared for the International Studies Association, February 23.

Adelman, I., and C. T. Morris (1973) *Economic Growth and Social Equity in Developing Countries.* Stanford: Stanford University Press.

Aldrich, H. E. (1979) *Organizations and Environments.* Englewood Cliffs, N.J.: Prentice-Hall.

Aldrich, H. E., and D. A. Whetten (1981) "Organization Sets, Action-Sets, and Networks: Making the Most of Simplicity." In Nystrom, P. C., and W. H. Starbuck (eds.) *Handbook of Organizational Design,* 1:385–408. New York: Oxford University Press.

Amber, G. H., and P. S. Amber (1962) *Anatomy of Automation.* Englewood Cliffs, N.J.: Prentice-Hall.

Anderson, R. S., and W. Huber (1988) *The Hour of The Fox: Tropical Forests, the World Bank, and Indigenous People in Central India.* Seattle: University of Washington Press.

Annis, S. (1986) "The Shifting Grounds of Poverty Lending at the World Bank." In Feinberg, R. E., and G. E. Helleiner (eds.) *Between Two Worlds: The World Bank's Next Decade,* 87–109. New Brunswick: Transaction Books.

Annis, S. (1987) "The Next World Bank: Can It Finance Development From the Bottom-Up." Paper presented to the Sub-Committee on International Development Institutions and Finance, U.S. House of Representatives, U.S. Congress, April 23.

Arnold, J. E. M. (1987) "Community Forestry." *Ambio* 16:122–128.

Arnold, J. E. M. (1988) Personal Communication. Letter dated April 18.

Arnold, J. E. M. (1992) *Community Forestry: Ten Years in Review.* FTP Community Forestry Note 7. Rome: FAO.

Arnold, S. H. (1982) *Implementing Development Assistance: European Approaches to Basic Needs.* Boulder: Westview Press.

Ashby, W. R. (1968) "Principles of the Self-Organizing System." In Buckley, W. (ed.) *Modern Systems Research for the Behavioral Scientist,* 108–118. Chicago: Aldine.

Astley, W. G. (1985) "The Two Ecologies: Population and Community Perspectives on Organizational Evaluation." *Administrative Science Quarterly* 30: 224– 241.

Astley, W. G., and A. H. Van de Ven (1983) "Central Perspectives and Debates in Organization Theory," *Administrative Science Quarterly* 28:245–273.

Ayres, R. L. (1983) *Banking on the Poor.* Cambridge: MIT Press.

Baum, W. C., and S. M. Tolbert (1985) *Investing in Development: Lessons of World Bank Experience.* Oxford: Oxford University Press.

Beckman, D. (1991) "Recent Experience and Emerging Trends." In Paul, S., and A. Israel (eds.) *Nongovernmental Organizations and the World Bank.* Washington, D.C.: World Bank.

Benson, J. K. (1975) "The Interorganizational Network as a Political Economy." *Administrative Science Quarterly* 20:229–249.

Bentley, W. R. (1993) "Essential Concepts of Agroforestry as Practiced in South Asia." In Bentley, W. R., P. K., Khosla, and K. Seckler (eds.) *Agroforestry in South Asia: Problems and Applied Research Perspectives.* New York: International Science Publisher.

Berg, R. J., and D. F. Gordon (eds.) (1989) *Cooperation for International Development: The United States and the Third World in the 1990s.* Boulder: Lynne Rienner Publishers.

Bertalanffy, L. von (1956) "General System Theory." In Bertalanffy, L. von, and A. Rapoport (eds.) *General Systems: Yearbook of the Society for the Advancement of General Systems Theory,* 1–10.

Blair, H. W. (1986) "Social Forestry: Time to Modify Goals?" *Economic and Political Weekly* 21:1317–1321.

Blair, H., P. May, and P. Olpadwala (1983) "Rural Institutions for the Development of Appropriate Forestry Enterprises." Cornell University, Rural Development Committee, Center for International Studies.

Blau, P. M., and R. A. Schoenherr (1971) *The Structure of Organizations.* New York: Basic Books.

Blau, P. M., and W. R. Scott (1962) *Formal Organizations.* San Francisco: Chandler.

Boardman, R. (1981) *International Organization and the Conservation of Nature.* Bloomington: Indiana University Press.

Braatz, S. (1985) *The Role of Development Assistance in Forestry: The Forestry Policies and Programs of the World Bank, the U.S. Agency for International Development, and the Canadian International Development Agency.* Washington, D.C.: International Institute for Environment and Development.

Brechin, S. R. (1989) "International Organization and Trees for People: A Sociological Analysis of the World Bank, FAO, CARE International, and Their Work and Performance in Community Forestry." Ph.D. dissertation, University of Michigan, Ann Arbor.

Brechin, S. R. (1994) "Forestry Project Data for the World Bank, FAO, and CARE International, 1969–1992." Mimeograph.

Brechin, S. R., and P. C. West (1982) "Social Barriers in Implementing Appropriate Techonology: The Case of Community Forestry in Niger, West Africa." *Humboldt Journal of Social Relations* 9:81–99.

Brechin, S. R., and P. C. West (1990) "Protected Areas, Resident Peoples and Sustainable Conservation: The Need to Link Top-Down with Bottom-Up." *Society and Natural Resources* 3:77–79.

Broad, R., and J. Cavanaugh (1993) *The Struggle for the Environment in the Philippines.* Berkeley: University of California Press.

Buckley, W. (1967) *Sociology and Modern Systems Theory.* Englewood Cliffs, N.J.: Prentice-Hall.

Buckman, R. E. (1987) "Strengthening Forestry Institutions in the Developing World", *Ambio* 16:120–121.

Burch, W. R., Jr. (1971) *Daydreams and Nightmares: A Sociological Essay of the American Environment.* New York: Harper and Row.

Burley, J. (1980) "Choice of Tree Species and Possibility of Genetic Improvement for Smallholder and Community Forestry." *Commonwealth Forestry Review* 59: 311–326.

Burley, J. (1982) "Obstacles to Tree Planting in Arid and Semi-Arid Lands: Comparative Case Studies from India and Kenya." Tokyo: United Nations University.

Burns, R. S., and G. M. Stalker (1961) *The Management of Innovation.* London: Tavistock.

Burt, R. S. (1980) "Models of Network Structure." *Annual Review of Sociology* 6: 79–141.

Caldwell, L. K. (1990) *International Environmental Policy: Emergence and Dimensions.* 2d ed. Durham: Duke University Press.

Campbell, J. L. (1994) "Recent Trends in Institutional Analysis: Bringing Culture Back into Political Economy." Paper presented at the Stanford Center for Organizations Research, Stanford University.

Campbell, W. J. (1990) *The History of CARE: A Personal Account.* New York: Praeger.

Carr-Harris, J. (1985) "Increasing NGO Involvement in Forestry: Some Implications from Senegal." *Unasylva* 37:26–31.

Cassen, R., and Associates (1986) *Does Aid Work? Report to an Intergovernmental Task Force.* Oxford: Clarendon Press.

CATIE (1979) *Agro-Forestry Systems in Latin America.* Costa Rica: CATIE.

Cazier, S. O. (1964) "CARE: A Study in Cooperative Voluntary Relief." Ph.D. dissertation, University of Wisconsin, Madison.

Cerena, M. (1985) "Alternative Units of Social Organization Sustaining Afforestation Strategies." In Cerena, M. (ed.) *Putting People First,* 267–293. New York: Oxford University Press.

Chadenet, B., and J. King (1972) "What Is a World Bank Project?" *Finance and Development* 9:2–12.

Chandrasekharan, C. (1983) "Rural Organizations in Forestry." *Unasylva* 35:2–11.

Child, J. (1972) "Organizational Structure, Environment and Performance: The Role of Strategic Choice." *Sociology* 6:2–22.

Child, J. (1975) "Managerial and Organizational Factors Associated with Company Performance, Part II: A Contingency Analysis." *Journal of Managerial Studies* 12:12–27.

Clements, P. (1993) "An Approach to Poverty Alleviation for Large International Development Agencies." *World Development* 21:1633–1646.

Conway, F. J. (1987) "Case Study: The Agroforestry Outreach Project in Haiti." Paper presented to the International Institute for Environment and Development's Conference on Sustainable Development, London.

Crane, B. B., and J. L. Finkle (1981) "Organizational Impediments to Development Assistance: The World Bank's Population Program." *World Politics* 33: 516–533.

CSE (1985) *The State of India's Environment, 1984–85: The Second Citizens' Report.* New Delhi.

DANIDA (1991) *Effectiveness of Multilateral Agencies at Country Level: Case Study of 2 Agencies in Kenya, Nepal, Sudan and Thailand.* Copenhagen: DANIDA, Ministry of Foreign Affairs.

Delehanty, J., M. Hoskins, and J. Thomson (1985) *Majjia Valley Evaluation Study: Sociological Report.* New York: Consultants Report to CARE.

De Montalembert, M. R. (1991) "Key Forestry Policy Issues in the Early 1990s." *Unasylva* 166:9–18.

DeRoy, R., and T. Mathew (1988) "Professional Profiles in Social Forestry." In Tree Project Clearing House (ed.) *The India Papers: Aspects of NGO Participation in Social Forestry,* 235–341. New York: Non-Governmental Liaison Office, United Nations.

DeSombre, E. R., and J. Kauffman (1994) "Montreal Protocol Multilateral Fund." In *The Effectiveness of Financial Transfer Mechanisms for International Environmental Protection.* Cambridge: Harvard University.

Dill, W. R. (1958) "Environment as an Influence on Managerial Autonomy." *Administrative Science Quarterly* 2:409–403.

DiMaggio, P. J., and W. W. Powell (1983) "The Iron Cage Revisited: Institutional Isomorphism and Collective Rationality in Organizational Fields." *American Sociological Review* 48:147–160.

DiMaggio, P. J., and W. W. Powell (1991) "Introduction." In Powell, W. W., and P. J. DiMaggio (eds.) *The New Institutionalism in Organizational Analysis,* 1–40. Chicago: University of Chicago Press.

Dobbin, F. (1994a) *Forging Industrial Policy: The United States, Britain and France in the Railway Age.* New York: Cambridge University Press.

Dobbin, F. (1994b) "Cultural Models of Organization: The Social Construction of Rational Organizing Principles." In Crane, D. (ed.) *The Sociology of Culture: Emerging Theoretical Perspectives.* 117–141. Oxford: Basil Blackwell.

Dornbusch, S. M., and R. W. Scott, with the assistance of B. C. Busching and J. D. Laing (1975) *Evaluation and the Exercise of Authority.* San Francisco: Jossey-Bass.

Easterbrook, G. (1995) *A Moment on the Earth: The Coming Age of Environmental Optimism.* New York: Viking.

Eckholm, E. (1979) *Planting for the Future: Forestry for Human Needs.* World Watch Paper 26. Washington, D.C.: World Watch Institute.

Eckholm, E. (1982) *Down To Earth.* New York: W. W. Norton.

El Razek, A. A. (1982) "International Refugee Assistance: A Study of the Determinants of Organizational Domain in the IRO and UNHCR." Ph.D. dissertation, University of Michigan.

Emerson, R. M. (1962) "Power-Dependence Relations." *American Sociological Review* 27:31–40.

Emery, F. E., and E. L. Trist (1965) "The Causal Texture of Organizational Environments." *Human Relations* 18:21–32.

Evans, W. M. (1966) "The Organizational Set: Toward a Theory of Interorganizational Relations." In Thompson, J. D. (ed.) *Approaches to Organizational Design,* 1–10. Pittsburgh: University of Pittsburgh Press.

Fairman, D., and M. Ross (1994) "International Aid for the Environment: Lessons from Economic Development Assistance." Working Paper Series, no. 94–4. The Center for International Affairs, Harvard University.

Feinberg, R. and G. Helleiner (1986) *Between Two Worlds: The World Bank's Next Decade.* New Brunswick: Transaction Books.

Finnemore, M. (1993) "International Organizations as Teachers of Norms: The United Nations Educational Scientific and Cultural Organization and Science Policy." *International Organization* 47:565–597.

Fisher, J. (1993) *The Road From Rio: Sustainable Development and the Nongovernmental Movement in the Third World.* Westport, CO: Praeger.

Foley, G., and G. Barnard (1984) *Farm and Community Forestry.* London: Earthscan.

Fontaine, R. G. (1985) "Forty Years of Forestry at FAO: Some Personal Reflections." *Unasylva* 37:5–14.

Fortmann, L. (1986). "Women in Subsistence Forestry: Cultural Myths Form a Stumbling Block." *Journal of Forestry* 10:39–42.

Fortmann, L. (1988) "Great Planting Disasters: Pitfalls in Technical Assistance in Forestry." *Agriculture and Human Values* 5:49–60.

French, D. (1985) "Monitoring and Evaluation of the Malawi Wood Energy Project." In *Monitoring and Evaluation of Participatory Forestry Projects.* 103–133. Rome: FAO.

Galbraith, J. (1973) *Designing Complex Organizations.* Reading, MA: Addison-Wesley.

Galbraith, J. (1977) *Organization Design.* Reading, MA: Addison-Wesley.

Ghai, D., and J. M. Vivian (eds.) *Grassroots Environmental Action: Peoples Participation in Sustainable Development.* New York: Routledge.

Gilmour, D. A., and R. J. Fisher (1991) *Villagers, Forests and Foresters: The Philosophy, Process and Practice of Community Forestry in Nepal.* Kathmandu, Nepal: Sahayogi Press.

Glassman, R. (1973) "Persistence and Loose Coupling in Living Systems." *Behavioral Science* 18:83–98.

Grainger, A. (1984) "Increasing the Effectiveness of Afforestation Projects in the Tropics Involving Non-Governmental Organizations." *International Tree Crop Journal* 3:33–47.

Gregerson, H., S. Draper, and D. Elz (eds.) (1989) "People and Trees: The Role of Social Forestry in Sustainable Development." Washington, D.C.: The World Bank.

Gupta, P. P., M. W. Dirsmith, and T. J. Fogarty (1994) "Coordination and Control in a Governmental Agency: Contingency and Institutional Theory Perspectives on GAO Audits." *Administrative Science Quarterly* 39:264–284.

Haas, E. (1990) *When Knowledge Is Power: Three Models of Change in International Organization.* Berkeley: University of California Press.

Haas, J. E., and T. E. Drabek (1973) *Complex Organizations: A Sociological Perspective.* New York: Macmillan.

Haas, P. M. (1989) *Saving the Mediterranean: The Politics of International Environmental Cooperation.* New York: Columbia University Press.

Haas, P. M. (ed.) (1992) "Knowledge, Power, and International Policy Coordination." *International Organization* 46, Special Issue.

Haas, P. M., M. A. Levy, and E. A. Parson (1992) "Appraising the Earth Summit: How Should We Judge UNCEN'S Success?" *Environment* 34:6–12.

Hage, J., and M. Aiken (1969) "Routine Technology, Social Structure, and Organization Goals." *Administrative Science Quarterly* 14:366–376.

Hage, J., and K. Finsterbusch (1987) *Organizational Change as a Development Strategy: Models and Tactics for Improving Third World Organizations.* Boulder: Lynne Rienner Publishers.

Hall, P. A. (1993) "Policy Paradigms, Social Learning, and the State: The Case of Economic Policymaking in Britain." *Comparative Politics* 4:275–296.

Hall, P. A., and R. C. R. Taylor (1994) "Political Science and the Four New Institutionalisms." Paper prepared for Annual Meeting of the American Political Science Association, Center for European Studies, Harvard University.

Hannan, M. T., and J. F. Freeman (1977) "The Population Ecology of Organizations." *American Journal of Sociology* 82:929–964.

Hannan, M. T., and J. F. Freeman (1984) "Structural Inertia and Organizational Change." *American Sociological Review* 49:149–164.

Hannan, M. T., and J. F. Freeman (1989) *Population Ecology.* Cambridge: Harvard University Press.

Harvey, E. (1968) "Technology and the Structure of Organizations." *American Sociological Review* 33:247–259.

Hawkins, E. K. (1970) *The Principles of Development Aid.* Baltimore: Penguin Books.

Hawley, A. (1950) *Human Ecology.* New York: Ronald Press.

Hayter, T. (1971) *Aid as Imperialism.* London: Penguin.

Hazlewood, P. (1987) "The Tropical Forestry Action Plan: Grass Roots Participation and the Role of Non-Governmental Organizations." Background paper, Bellagio strategy meeting on tropical forests, Bellagio Study and Conference Center, Lake Como, Italy, July 12.

Heermans, J. (1987) "An Evaluation of the World Bank's Irrigated Energy Project in Niger, West Africa." Report to USAID Mission, Niamey, Niger.

Hill, M. (1978) *The United Nations System: Coordinating Its Economic and Social Work.* Cambridge: Cambridge University Press.

Hirsch, P. M. (1985) "The Study of Industries." In Bacharach, S. B., and S. M. Mitchell (eds.) *Research in the Sociology of Organizations,* 4:271–309. Greenwich, Conn. JAI Press.

Hisham, M. A., J. Sharma, A. Ngaiza, and N. Atampugre (1991) *Whose Trees? A People's View of Foerstry Aid.* London: The Panos Institute.

Hobens, A. (1989) "USAID: Organizational and Institutional Issues and Effectiveness." In Berg, R. J., and D. F. Gordon (eds.) *Cooperation for International Development: The United States and the Third World in 1990s.* Boulder: Westview Press.

Homans, G. C. (1949) "The Strategy of Industrial Sociology," *American Journal of Sociology* 54:330–337.

Hoskins, M. (1980) "Community Participation in Africa Fuelwood Production, Transformation, and Utilization," In French, D., and L. Larson (eds.) *Energy for Africa: Selected Readings,* 155–188. Washington, D.C.: USAID.

Hoskins, M. (1980) "Social Dimensions in Local Forestry/Conservation Efforts." Paper for sociological workshop on forestry projects, World Bank, October.

Hoskins, M. (1983) "Mobilizing Rural Communities." *Unaslva* 35:12–15.

Hurni, B. S. (1980) *The Lending Policy of the World Bank in the 1970s: Analysis and Evaluation.* Boulder: Westview Press.

Huxley, P. A. (1983) *Plant Research and Agroforestry.* Nairobi: ICRAF.

Hyman, E. L. (1983) "Small Tree Farming in the Philippines: A Comparison of Two Credit Programs." *Unasylva* 35:25–33.

Ickis, J. C., E. de Jesus, and R. Maru (1986) *Beyond Bureaucracy: Strategic Management of Social Development.* Hartford, Conn. Kumarian Press.

ICRAF (1986) *Global Inventory of Agroforestry Systems.* Nairobi: ICRAF.

Singh, S. (1987) *The India Papers: Aspects of NGO Participation in Social Forestry.* New York: International Tree Project Clearinghouse, Non-Government Liaison Service, United Nations, in association with the NGO Steering Committee of the Bangalore Consultation.

Israel, A. (1987) *Institutional Development: Incentives to Performance.* Published for The World Bank. Baltimore: Johns Hopkins University Press.

Jacobson, Harold K. (1984) *Networks of Interdependence.* New York: Knopf.

Jay, K., and C. Michalopoulos (1989) "Donor Policies, Donor Interests and Aid Effectiveness." In Krueger, A., C. Michalopoulos, and V. Ruttan (eds.) *Aid and Development,* 68–88. Baltimore: Johns Hopkins University Press.

Johnston, B. R. (ed.) (1994) *Who Pays the Price? The Sociocultural Context of Environmental Crisis.* Washington, D.C.: Island Press.

Kahn, R. L., and M. N. Zald (eds.) (1990) *Organizations and Nation-States: New Perspectives on Conflict and Cooperation.* San Francisco: Jossey-Bass.

Katz, D., and R. L. Kahn (1966) *The Social Psychology of Organizations.* New York: Wiley.

Katz, H. (1974) *Give! Who Gets Your Charity Dollars.* Garden City, New York: Anchor Press/Doubleday.

Kaufman, H. (1960) *The Forest Ranger.* Baltimore: Johns Hopkins University Press.

Kaufman, D. (1971) *A History of CARE.* New York: CARE Archives.

Kay, D. (1980) *The Functioning and Effectiveness of Selected United Nations Programs,* Washington, D. C.: The American Society of International Law.

Kay, D., and H. Jacobson, (eds.) *Environmental Protection: The International Dimension.* New Jersey: Allanhel, Osmun.

Kendall, H. W., and D. Pimentel, (1994) "Constraints on the Expansion of the Global Food." *Amibo* 23:3.

King, J. A. (1974) "Reorganizing the World Bank." *Finance and Development* 11:5–8.

Korten, D. C. (1980) "Community Organization and Rural Development: A Learning Process Approach." *Public Administrative Review* 40:480–511.

Korton, D. C. (ed.) (1987) *Community Management: Asian Experience and Perspectives.* West Hartford, Conn. Kumarian Press.

Korten, F. F. (1994) "Questioning the Call for Environmental Loans: A Critical Examination of Forestry Lending in the Philippines." *World Development* 22: 971–981.

Korton, D., and N. T. Uphloff (1981) "Bureaucratic Reorientation for Participatory Rural Development." Ithaca, N.Y.: National Association of Schools of Public Affairs and Administration in association with Cornell University.

Krasner, S. D. (ed.) (1983) *International Regimes.* Ithaca, N.Y.: Cornell University Press.

Lawrence, P. R., and J. W. Lorsch (1967) *Organization and Environment: Managing Differentiation and Integration.* Boston: Graduate School of Business Administration, Harvard University.

Leach, G., and R. Mearns (1988) *Beyond The Woodfuel Crisis: People, Land and Trees in Africa.* London: Earthscan.

Leloup, M. (1985) "The First Ten Years." *Unasylva* 30:39–40.

LePrestre, P. G. (1985) "The Ecology of International Organizations." *International Interactions* 12:21–44.

Levine, S., and P. E. White (1961) "Exchange as a Conceptual Framework for the

Study of Interorganizational Relationships." *Administrative Science Quarterly* 5: 583–601.

Levy, M. A., O. R. Young, and M. Zurn (1994) "The Study of International Regimes." Working paper, IIASA, Laxenburg, Austria.

Levy, M. A., R. O. Keohane, and W. Clark (1994) *The Effectiveness of Financial Transfer Mechanisms for International Environmental Protection.* Background materials, Center for International Affairs and Center for Science and Intearnational Affairs, Harvard University.

Lewis, W. A. (1954) "Economic Development with Unlimited Supplies of Labor." *Manchester School of Economics and Social Studies* 22:139–191.

Litwak, E. (1961) "Models of Bureaucracy Which Permit Conflict." *American Journal of Sociology* 67:177–184.

Lohmann, L., and M. Colchester (1990) "Paved with Good Intentions: TFAP's Road To Oblivion." *The Ecologist,* 20:91–98.

Luard, E. (ed.) (1966) *The Evolution of International Organizations.* London: Thames and Hudson.

Lynch, B. P. (1974) "An Empirical Assessment of Perrow's Technology Construct." *Administrative Science Quarterly* 19:338–356.

Mason, E., and R. E. Asher (1973) *The World Bank Since Bretton Woods.* Washington, D.C.: The Brookings Institution.

McGaughey, S. E. (1986) "International Financing for Forestry." *Unasylva* 38:2–11.

Meyer, J. W., and B. Rowan (1977) "Institutionalized Organizations: Formal Structure as Myth and Ceremony." *American Journal of Sociology* 83:340–363.

Meyer, J. W., and W. R. Scott (1983) *Organizational Environments: Ritual and Rationality.* Beverly Hills: Sage.

Meyer, M. W. (1979) "Organizational Structure as Signaling." *Pacific Sociological Review* 22:481–500.

Mitchell, R. (1994) *International Oil Pollution at Sea.* Cambridge, MA: MIT Press.

Mitrany, D. (1966) *A Working Peace System.* Chicago: Quadrangle Books.

Muthoo, M. K. (1985) "FAO's Field Programme: The First 40 Years." *Unasylva* 37: 52–58.

Muthoo, M. K. (1991) "An Overview of the FAO Forestry Field Programme." *Unasylva* 42:30–39.

Myrdal, G. (1978) "In Memoriam: Egon Gleringer's Contribution to International Forestry and FAO." *Unasylva* 30:39–40.

National Academy of Sciences (1977) *Leucaena: Promising Forage and Tree Crops for the Future.* Washington, D.C.: National Academy of Sciences.

National Academy of Sciences (1979) *Tropical Leumes: Resources for the Future,* Washington, D.C.: National Academy of Sciences.

National Academy of Sciences (1980) *Firewood Crops: Shrubs and Tree Species for Energy Production.* Washington, D.C.: National Academy of Sciences.

Nelson, P. J. (1991) "The World Bank and Non-Governmental Organizations: Political Economy and Organizational Analysis." Ph.D. dissertation, University of Wisconsin, Madison.

Ness, G. D., and S. R. Brechin (1988) "Bridging the Gap: International Organizations as Organizations." *International Organization* 42:245–273.

Noronha, R., and J. Spears (1985) "Sociological Variables in Forestry Project Design." In Cernea, M. (ed.) *Putting People First: Sociological Variables in Rural Development,* 227–266. New York: Oxford University Press.

Oksanen, T., M. Heering, and B. Cabarle (1993) *A Study of Coordination in Sustainable Forestry Development.* Washington, D.C.: TFAP Forestry Advisers Group.

Paul, S. P., and A. Israel (1991) *Nongovernmental Organizations and the World Bank: Cooperation for Development.* Washington, D.C.: World Bank.

Peluso, N. L., M. Poffenberger, and F. Seymour (1990) "Reorienting Forest Management on Java." In Poffenberger, M. (ed.) *Keepers of the Forest: Land Management Alternatives in Southeast Asia,* 220–236. West Hartford, Conn. Kumarian Press.

Pennings, J. M. (1992) "Structural Contingency Theory: A Reappraisal." In Straw, B. M. and L. L. Cummings (eds.) *Research in Organizational Behavior: An Annual Series of Analytical Essays and Critical Reviews.* Vol. Greenwich, Conn. JAI Press.

Perrow, C. (1967) "A Framework for the Comparative Analysis of Organizations." *American Sociological Review* 32:194–208.

Perrow, C. (1970) *Organizational Analysis: A Sociological View.* Belmont, Calif.: Wadsworth.

Perrow, C. (1984) *Normal Accidents: Living with High-Risk Technologies.* New York: Basic Books.

Perrow, C. (1986) *Complex Organizations: A Critical Essay.* 3d ed. New York: Random House.

Pfeffer, J., and G. R. Salancik (1978) *The External Control of Organizations.* New York: Harper and Row.

Phaup, E. D. (1984) *The World Bank: How It Can Serve U.S. Interests.* Washington, D.C.: Heritage Foundation.

Phillips, R. W. (1981) *FAO: Its Origins, Formation and Evaluation,* 1945–1981. Rome: FAO.

Poffenberger, M. (1990) *Keepers of the Forest: Land Management Alternatives in Southeast Asia,* West Hartford, Conn. Kumarian Press.

Postel, S., and L. Heise (1988) *Reforesting the Earth.* Washington, D.C.: Worldwatch Institute.

Powell, W. W., and P. J. Di Maggio (eds.) *The New Institutionalism in Organizational Analysis.* Chicago: University of Chicago Press.

Pugh, D. S., D. J. Hickson, C. R. Hinings, and C. Turner (1969) "The Context of Organization Structures." *Administrative Sciences Quarterly* 14:91–114.

Rackham, J., and J. Woodward (1970) "The Measurement of Technical Variables."

In Woodward, J. (ed.) *Industrial Organization: Behavior and Control.* 19–36. London: Oxford University Press.

Raintree, J. B. (ed.) (1987) *Land, Trees and Tenure: Proceedings of an International Workshop on Tenure Issues in Agroforestry.* Madison and Nairobi: Land Tenure Center and the International Council for Research in Agroforestry.

Ravak, S. R. (1988) "Social Forestry and Resource Utilization: History and Prospects in India." Paper presented at the Second Symposium on Social Science in Resource Management Institute of Environmental Studies, University of Illinois, June 7–9.

Reiss, E. C. (1985) *The American Council of Voluntary Agencies for Foreign Service.* New York: American Council of Voluntary Agencies for Foreign Service.

Rich, B. (1986) "Environmental Management and Multilateral Development Banks." *Cultural Survival Quarterly* 10:4–13.

Rich, B. (1994) *Mortgaging the Earth: The World Bank, Environmental Impoverishment, and the Crisis of Development.* Boston: Beacon Press.

Riddell, R. C. (1987) *Foreign Aid Reconsidered.* Baltimore: Johns Hopkins University Press.

Rogers, D. L., D. A. Whetten, and Associates (1982) *Interorganizational Coordination: Theory, Research, and Implementation.* Ames: Iowa State University Press.

Rondinelli, D. A. (1987) *Development Administration and U.S. Foreign Aid Policy.* Boulder, CO: Lynne Rienner Publishers.

Rushing, W. A. (1968) "Hardness of Material as an External Constraint on the Division of Labor in Manufacturing Industries." *Administrative Science Quarterly* 13:229–245.

Saouma, E. (1991) "Forestry in the 1990s: An interview with FAO Director-General Edouard Saouma." *Unasylva* 42:3–8.

Saouma, E. (1993) *FAO in the Front Line of Development.* Rome: FAO.

Schiff, A. L. (1962) *Fire and Water: Scientific Heresy in Forest Service.* Cambridge: Harvard University Press.

Scott, W. R. (1981) *Organizations: Rational, Natural and Open Systems,* Englewood Cliffs, NJ: Prentice-Hall.

Scott, W. R., and J. W. Meyer (1983) "The Organization of Societal Sectors." In Meyer, J. W., and W. R. Scott (eds.) *Organizational Environments: Ritual and Rationality,* 129–154. Beverly Hills, Calif.: Sage Publications.

Scott, W. R. (1987.) "The Adolescence of Institutional Theory." *Administrative Science Quarterly* 32:493–511.

Scott, W. R. (1987) *Organizations: Rational, Natural and Open Systems.* 2d ed. Englewood Cliffs, N.J.: Prentice-Hall.

Scott, W. R. (1992) *Organizations: Rational, Natural and Open Systems,* 3d ed. Englewood Cliffs, N.J.: Prentice-Hall.

Selznick, P. (1949) *TVA and the Grass Roots: A Study of Politics and Organizations,* Berkeley: University of California Press.

Selznick, P. (1957) *Leadership in Administration: A Sociological Interpretation.* Berkeley: University of California Press.

Shepherd, G. (1992) *Managing Africa's Tropical Dry Forests: A Review of Indigenous Methods.* ODI Agricultural Occasional Paper 14. London: Overseas Development Institute.

Shah, S. A. (1987) "Tree Planting for Rural People Involvement of Non-Governmental Organizations." *The International Tree Crops Journal* 4:195–207.

Shiva, V. (1987) "Forestry Myths and the World Bank. A Cultural Review of Tropical Forests: A Call for Action." *The Ecologist* 17:142–149.

Shiva, V., H. C. Sharatchandra, and J. Bandyopadhyay (1987) "Social Forestry for Whom?" In Korten, D. C. (ed.) *Community Management: Asian Experience and Perspectives,* 238–246. West Hartford, Conn. Kumarian Press.

Shiva, V. (1991) *Ecology and the Politics of Survival: Conflicts over, Natural Resources in India.* London: United Nations University Press.

Sierra Club (1986) *Bankrolling Disasters: International Development Banks and the Global Environment.* Washington, D.C.: Sierra Club.

Sklair, L. (1991) *Sociology of the Global System.* New York: Harvester Wheatsheaf.

Stein, R. E., and B. D. Johnson (1979) *Banking on the Biosphere.* Lexington: Lexington Books.

Strang, D., and J. W. Meyer (1993) "Institutional Conditions for Diffusion." *Theory and Society* 22:487–511.

Strang D., and P. M. Y. Chang (1993) "The International Labor Organisation and the Welfare State: Institutional Effects on National Welfare Spending, 1960–80." *International Organization* 47:235–262.

TAICH (1983) *U.S. Non-Profit Organizations in Development Assistance Abroad.* New York: TAICH.

Tapp, C., T. Resh, L. Bush, and L. Ntiu (1986) "Uganda Village Forestry Project." New York: CARE.

Tendler, J. (1975) *Inside Foreign Aid.* Baltimore: Johns Hopkins University Press.

Tendler, J. (1982) "Turning Private Voluntary Organizations into Development Agencies: Questions for Evaluations." A.I.D. Program Evaluation Discussion Paper No. 12. Washington, D.C.

Terreberry, S. (1968) "The Evolution of Organizational Environments." *Administrative Science Quarterly* 12:590–613.

Thompson, J. D., and F. L. Bates (1957) "Technology, Organization, and Administration." *Administrative Science Quarterly* 2:325–342.

Thompson, J. D. (1967) *Organizations in Action.* New York: McGraw-Hill.

Thomson, J. T. (1977) "Ecological Deterioration: Local Level Rule-Making and Enforcement Problems in Niger." In Glantz, M. (ed.) *Desertification: Environmental Degradation in and Around Arid Lands,* 57–79. Boulder Colo.: Westview Press.

Thomson, J. T. (1981) "Rules, Trees, and Reforestation in the Nigerian Sahel." *Sylva Africana* 10:5–6.

Udy, S. H., Jr. (1979) *Organization of Work.* New Haven, Conn.: Human Relations Area Files Press.

Van de Laar, A. H. (1980) *The World Bank and the Poor.* Boston: Martinus Nijhoff.

Van de Ven, A. H., and A. Delbecq (1974) "A Task Contingent Model of Work-Unit Structure." *Administrative Science Quarterly* 19:183–197.

Van Wicklin, W. A. (1990) "Private Voluntary Organization as Agents of Alternative Development Strategies." Ph.D. dissertation, Massachusetts Institute of Technology.

Waltz, K. N. (1979) *Theory of International Politics.* New York: Random House.

Wamsley, G. L., and M. N. Zald (1973) *The Political Economy of Public Organizations,* Lexington, Mass.: Heath.

Warren, R. L. (1967) "The Interorganizational Field as a Focus for Investigation." *Administrative Science Quarterly* 12:396–419.

Washington Post (1981) "Arthur Ringland, 99, Conservationist, Whose Efforts Lead to the Founding of CARE." Wednesday, October 21, p. C10.

Weber, F. (1985) "Haiti Agroforestry Outreach Project, CARE Subproject: Assessment of Research Needs and Modalities." Consultants Report for CARE, NY.

West, P. C. (1982) *Natural Resource Bureaucracy and Rural Poverty.* Ann Arbor, Mich. Natural Resource Sociology Research Lab.

West, P. C. (1983) "Collective Adoption of Natural Resource Practices in Developing Nations." *Rural Sociology* 48:44–59.

West, P. C. (1984) "Sociological Aspects of Agroforestry." In Shapiro, K. (ed.) *Agroforestry in Developing Countries,* 33–71. Ann Arbor: Center for Research on Economics Development, University of Michigan.

West, P. C. (1994) "Natural Resources and the Persistence of Rural Poverty in America: A Weberian Perspective on the Role of Power, Domination, and Natural Resource Bureaucracy." *Society and Natural Resources* 7:415–427.

Westoby, J. C. (1975) "Making Trees Serve People." *The Commonwealth Forestry Review* 54:206–215.

Westoby, J. C. (1987) *The Purpose of Forests: Follies of Development.* New York: Basil Blackwell.

Westoby, J. C. (1988) Personal communication, letter dated March 31.

Westoby, J. C. (1989) *Introduction to World Forestry: People and Their Trees.* New York: Basil Blackwell.

Wiersum, K. F. (1984) *Strategies and Designs for Afforestation, Reforestation and Tree Planting.* Proceedings of an International Symposium. Wageningen, Netherlands: PUDOC.

Winterbottom, R. (1980) "Reforestation in the Sahel: Problems and Strategies." Paper presented at the African Studies Association, Philadelphia.

Winterbottom, R., and P. T. Hazelwood (1987) "Agroforestry and Sustainable Development: Making the Connection." *Ambio* 16:100–110.

Winterbottom, R. (1990) "Taking Stock: The Tropical Forestry Action Plan After Five Years." Washington, D.C.: World Resources Institute.

Winters, R. K. (1974) *The Forest and Man.* New York: Vantage Press.

Woodward, J. (1958) *Management and Technology.* London: H.M.S.O.

Woodward, J. (1965) *Industrial Organization: Theory and Practice.* New York: Oxford University Press.

Woodward, J. (ed.) (1970) *Industrial Organization: Behavior and Control.* London: Oxford University Press.

World Resources Institute (1985) *Tropical Forests: A Call for Action, Part I: The Plan.* Washington, D.C.: World Resources Institute.

World Resources Institute (1987) *World Resources, 1987.* Washington, D.C.: World Resources Institute.

Young, O. R. (1989) *International Cooperation: Building Regimes for Natural Resources and the Environment.* Ithaca, NY: Cornell University Press.

Yudelman, M. (1985) "The World Bank and Agricultural Development: An Insider's View." *WRI Papers,* vol. 1. Washington, D.C.

Zald, M. N. (1970) "Political Economy: A Framework for Comparative Analyses." In Zald, M. N. (ed.) *Power in Organizations,* 221–261. Nashville: Vanderbilt University Press.

Organizational Documents

CARE (1947) *1947 Annual Report.* New York.

CARE (1956) *1956 Annual Report.* New York.

CARE (1961) *1961 Annual Report.* New York.

CARE (1966) *1966 Annual Report.* New York.

CARE (1971–78) *Annual Reports.* New York.

CARE (1979a) *1979 Annual Report.* New York.

CARE (1979b) *Care Overseas Operation Manual.* New York.

CARE (1980–84) *Annual Reports.* New York.

CARE (1985a) *1985 Annual Report.* New York.

CARE (1985b) "Rural Capital Formation." Grant proposal to USAID, New York.

CARE (1985c) *Renewable Natural Resources: CARE/USAID Matching Grant Final Report.* New York.

CARE (1986a) *CARE: A Profile.* New York.

CARE (1986b) *1986 Annual Report.* New York.

CARE (1986c) *The Maiiia Valley Windbreak Evaluation Study: An Examination of Progress and Analyses to Date.* Field report prepared by Steve Dennison.

CARE (1987a) "World Bank Meeting Agenda." Internal CARE document, January 23.

CARE (1987b) *1987 Annual Report.* New York.

CARE (1988–94) *Annual Reports.* New York.

Chemonics International (1983) *Review of the CARE Guatemala Reforestation and Soil Conservation Project Final Report.* Consultants report to CARE, New York.

COFO (1972a) *Forestry and the Environment. Action Proposals on Forestry, National Parks, and Wildlife for the UN Conference on the Human Environment: Secretariat Note.* Committee on Forestry, First Session, Rome.

COFO (1972b) *United Nations Conference on the Human Environment.* Committee on Forestry, Supplemental 1, Rome.

COFO (1974a) *FAO Medium-Term Objectives and Proposals for the Forestry Department's Programme of Work, 1976–77: Secretariat Note.* Committee on Forestry, Second Session, Rome.

COFO (1974b) *Field Activities of the Forestry Department: Secretariat Note.* Committee on Forestry, Second Session, Rome.

COFO (1974c) *The Role and Participation of Forestry in the World Endeavour for the Conservation of the Environment: Secretariat Note.* Committee on Forestry, Second Session, Rome.

COFO (1974d) *The Rules of Procedure of the Committee on Forestry: Secretariat Note.* Committee on Forestry, Second Session, Rome.

COFO (1974e) *The Role of the Regional Forestry Commissions: Secretariat Note.* Committee on Forestry, Second Session, Rome.

COFO (1974f) *The Eighth World Forestry Congress: Secretariat Note.* Committee on Forestry, Second Session, Rome.

COFO (1976a) *Forestry for Local Community Development: Secretariat Note.* Committee on Forestry, Third Session, Rome.

COFO (1976b) *Future Programmes of Work of the Forestry Department: Secretariat Note.* Committee on Forestry, Third Session, Rome.

COFO (1976c) *Review of Forestry Field Programmes: Secretariat Note.* Committee on Forestry, Third Session, Rome.

COFO (1978a) *Definition of Terms for Use in Forestry for Local Community Development: Secretariat Note.* Committee on Forestry, Fourth Session, Rome.

COFO (1978b) *Development and Investment in the Forestry Sector: Secretariat Note.* Committee on Forestry, Fourth Session, Rome.

COFO (1978c) *Eighth World Forestry Congress: Secretariat Note.* Committee on Forestry, Fourth Session, Rome.

COFO (1978d) *Review of Forestry Field Programmes: Secretariat Note.* Committee on Forestry, Fourth Session, Rome.

COFO (1978e) *Review of Progress Made in 1976–77 and Main Features of the Programme of Work in Forestry for 1978–79: Secretariat Note.* Committee on Forestry, Fourth Session, Rome.

COFO (1978f) *Small-Scale Forest Industries for Development: Secretariat Note.* Committee on Forestry, Fourth Session, Rome.

COFO (1978g) *The Place of Forests and Trees in Integrated Rural Development (with Par-*

ticular Reference to Tropical Countries): Secretariat Note. Committee on Forestry, Fourth Session, Rome.

COFO (1980a) *FAO's Medium-Term Objective and Proposals for Future Programmes in Forestry: Secretariat Note.* Committee on Forestry, Fifth Session, Rome.

COFO (1980b) *Review of Forestry Field Programmes: Secretariat Note.* Committee on Forestry, Fifth Session, Rome.

COFO (1980c) *Review of Progress Made in 1978–79 and Main Features of the Programme of Work in Forestry for 1980–81: Secretariat Note.* Committee on Forestry, Fifth Session, Rome.

COFO (1980d) *Towards A Forestry Strategy For Development: Secretariat Note.* Committee on Forestry, Fifth Session, Rome.

COFO (1982a) *Review of Forestry Field Programmes: Secretariat Note.* Committee on Forestry, Sixth Session, Rome.

COFO (1982b) *Review of Progress Made in 1980–81 and Main Features of the Programme of Work in Forestry for 1982–83: Secretariat Note.* Committee on Forestry, Sixth Session, Rome.

COFO (1984a) *Implementation of the Forestry Development Strategy Adopted by the Committee at its Fifth Session (May 1980): Secretariat Note.* Committee on Forestry, Seventh Session, Rome.

COFO (1984b) *Review of Forestry Field Programmes: Secretariat Note.* Committee on Forestry, Seventh Session, Rome.

COFO (1986a) *Progress made in 1984–85 and Main Features of the Programme of Work in Forestry for 1986–87: Secretariat Note.* Committee on Forestry, Eighth Session, Rome.

COFO (1986b) *Review of Forestry Field Programmes, 1984–85: Secretariat Note.* Committee on Forestry, Eighth Session, Rome.

COFO (1986c) *Review of Progress Made in 1982–83 and Main Features of the Programme of Work in Forestry for 1984–85: Secretariat Note.* Committee on Forestry, Seventh Session, Rome.

FAO (1974) *Tree Planting Practices in African Savannas.* Rome: FAO.

FAO (1978) *Forestry for Local Community Development.* Rome: FAO.

FAO (1983) *Fighting Rural Poverty: FAO's Action Programme for Agrarian Reform and Rural Development.* Rome: FAO.

FAO (1984) *Basic Texts of the Food and Agriculture Organizations of the United Nations.* Vols. 1 and 2. Rome: FAO.

FAO (1985a) *FAO: The First 40 Years: 1945–85.* Rome: FAO.

FAO (1985b) *Forestry for Development.* Rome: FAO.

FAO (1985c) *Forests, Tress and People.* Forestry Topics Report No. 2. Rome: FAO.

FAO (1985d) *Tree Growing by Rural People.* FAO Forestry Paper No. 64. Rome: FAO.

FAO (1986a) *Forestry Extension Organization.* FAO Forestry Paper No. 66. Rome: FAO.

FAO (1986b) *FAO: What It Is. What It Does.* Rome: FAO.

FAO (1987) *Terminal Report of the Forestry for Local Community Development Programme.* Rome.
FAO (1993) *FAO Annual Review: A Summary of the Organization's Activities During 1993.* Rome.
FAO Conference (1961) *The Director-General's Programme of Work and Budget for 1962–63.* Rome.
FAO Conference (1963) *The Director-General's Programme of Work and Budget for 1964–65.* Rome.
FAO Conference (1965) *The Director-General's Programme of Work and Budget for 1966–67.* Rome.
FAO Conference (1967) *The Director-General's Programme of Work and Budget for 1968–69.* Rome.
FAO Conference (1969) *The Director-General's Programme of Work and Budget for 1970–71.* Rome.
FAO Conference (1971) *The Director-General's Programme of Work and Budget for 1972–73.* Rome.
FAO Conference (1973a) *The Director-General's Programme of Work and Budget for 1974–75.* Rome.
FAO Conference (1973b) *Medium-Term Objective.* Rome.
FAO Conference (1975) *The Director-General's Programme of Work and Budget for 1976–77.* Rome.
FAO Conference (1977) *The Director-General's Programme of Work and Budget for 1978–79.* Rome.
FAO Conference (1979a) *The Director-General's Programme of Work and Budget for 1980–81.* Rome.
FAO Conference (1979b) *Review of the Regular Program.* Rome.
FAO Conference (1981a) *The Director-General's Programme of Work and Budget for 1982–83.* Rome.
FAO Conference (1981b) *Review of the Regular Programme, 1980–81.* Rome.
FAO Conference (1983) *The Director-General's Programme of Work and Budget for 1984–85.* Rome.
FAO Conference (1983) *Review of the Regular Programme, 1982–83.* Rome.
FAO Conference (1985a) *The Director-General's Programme of Work and Budget for 1986–87.* Rome.
FAO Conference (1985b) *Review of the Regular Programme, 1984–85.* Rome.
FAO Conference (1985c) *Review of the Field Programme, 1984–85.* Rome.
FAO Conference (1987) *The Director-General's Programme of Work and Budget for 1988–89.* Rome.
FAO Council (1974) *Report of the Second Session of the Committee on Forestry.* Rome.
FAO Council (1988) *Report on the Ninth Session of the Committee on Forestry.* Rome.
FAO Council (1990) *Report on the Tenth Session of the Committee on Forestry.* Rome.
FAO Council (1993) *Report on the Eleventh Session of the Committee on Forestry.* Rome.

FAO Forestry (1977–86) "Forestry Planning Newsletter," nos. 1–10. Rome.

World Bank (1951–90) *World Bank Annual Report.* Washington, D.C.

World Bank (1978a) *Forestry.* Sector Policy Paper. Washington, D.C.: World Bank.

World Bank (1978–86) "Review of the World Bank Financed Forestry Activity." Washington, D.C. Mimeograph.

World Bank (1991a) *The Forestry Sector: A World Bank Policy Paper.* Washington, D.C.

World Bank (1991b) *Forestry: The World Bank's Experience.* Washington, D.C.

World Bank (1991c) *Environmental Assessment Sourcebook, Volume 1: Policies, Procedures, and Cross-Sectoral Issues.* Washington, D.C.

World Bank (1992a) *World Bank Annual Report, 1992.* Washington, D.C.

World Bank (1992b) *Effective Implementation: Key to Development Impact.* Washington, D.C.

World Bank (1994) "Review of Implementation of the Forest Sector Policy." Washington, D.C.

Index

About the Author

STEVEN R. BRECHIN, PH.D., is an associate research scientist at the International Institute of the University of Michigan, Ann Arbor, where he specializes in the integration of social science perspectives in both international and domestic environmental and natural resource concerns. Previously, he was assistant director of the Princeton Environmental Institute and a senior lecturer in the Woodrow Wilson School of Public and International Affairs at Princeton University. He is coeditor of *Population-Environment Dynamics: Ideas and Observations* (University of Michigan Press, 1993) and *Resident Peoples and National Parks: Social Dilemmas and Strategies in International Conservation* (University of Arizona Press, 1991). His articles have appeared in such peer-reviewed journals as *Social Science Quarterly, Society and Natural Resources,* and *International Organization.*

LIBRARY OF CONGRESS CATALOGING-IN-PUBLICATION DATA

Brechin, Steven R., 1953–
Planting trees in the developing world : a sociology of international organizations / Steven R. Brechin.
p. cm.
"Published in cooperation with the Center for American Places, Harrisonburg, Virginia"—T.p. verso.
Includes bibliographical references (p.) and index.
ISBN 0-8018-5439-3 (alk. paper)
1. Forestry projects—Developing countries—Sociological aspects. 2. Social forestry programs—Developing countries. 3. International agencies—Sociological aspects. I. Center for American Places (Harrisonburg, Va.) II. Title.
SD387.P74B74 1997
306.3'49—dc20 96-27636
CIP